ASTRONOMY FACTFINDER

Author
Tim Furniss

Editor
Steve Parker

Design
Penatcor

Image Coordination
Ian Paulyn

Production Assistant
Jenni Cozens

Index
Jane Parker

Editorial Director
Paula Borton

Design Director
Clare Sleven

Publishing Director
Jim Miles

This edition published by Dempsey Parr, 1999
Dempsey Parr, Queen Street House, 4 Queen Street, Bath, BA1 1HE, UK

2 4 6 8 10 9 7 5 3 1

Produced by Miles Kelly Publishing Ltd
Bardfield Centre, Great Bardfield, Essex CM7 4SL

ISBN 1 84084 518 X

Printed in Singapore

ASTRONOMY

FACTFINDER

Introduction

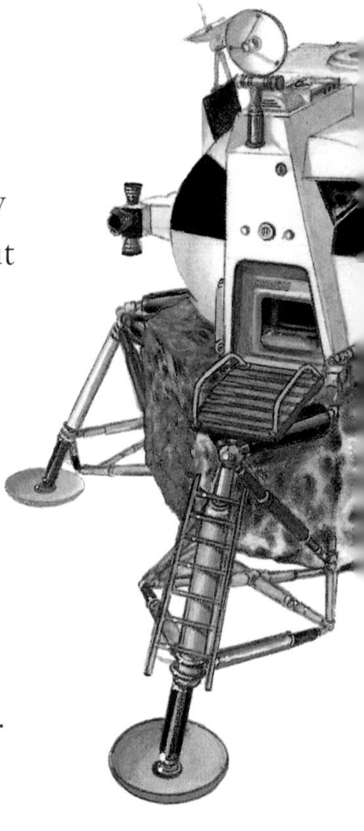

Humans are visual creatures. We rely on eyesight more than any other sense — especially to find out about the world around us from words and pictures. Lists of facts and descriptions of events may contain concentrated knowledge, but adding illustrations helps to bring the subject alive. They encourage us to delve further, appreciate, and enjoy, as well as to retain the information.

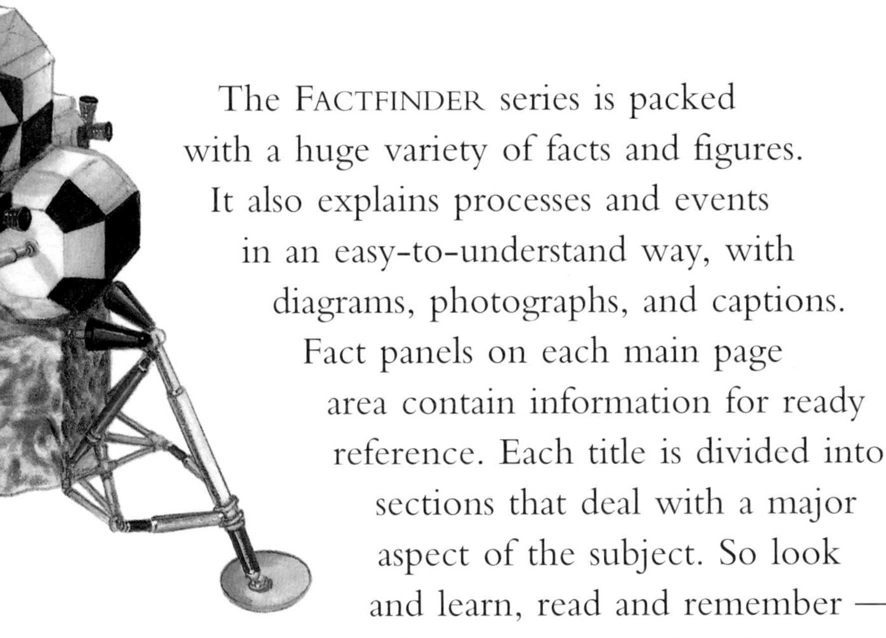

The FACTFINDER series is packed with a huge variety of facts and figures. It also explains processes and events in an easy-to-understand way, with diagrams, photographs, and captions. Fact panels on each main page area contain information for ready reference. Each title is divided into sections that deal with a major aspect of the subject. So look and learn, read and remember — and return again and again.

CONTENTS

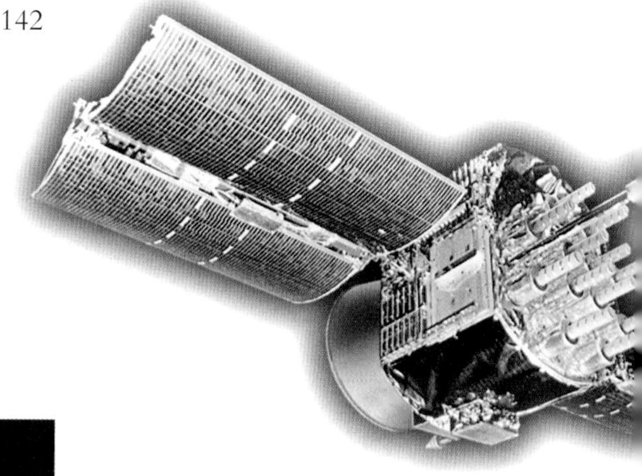

Our Earth

Ever since the first humans peered in wonder at the stars in the night sky, we have longed to know more about the Universe. Today, we have the technology to explore deeper into the Universe than ever imagined. The Earth is just one very tiny part of the vast Universe — a small rocky planet traveling around a medium-sized star, the Sun, in one of billions of galaxies. Nobody really knows where the Universe begins or ends. Even though the most powerful radio telescope has detected a quasar 13.2 billion light-years away, we have so far explored only a very small part of the Universe.

THE EARTH AND ITS MOON
This is a view of the Earth as seen from another world in space — from our neighbor, the Moon. This classic photograph of the space age was taken by the crew of Apollo *8 in December 1968.*

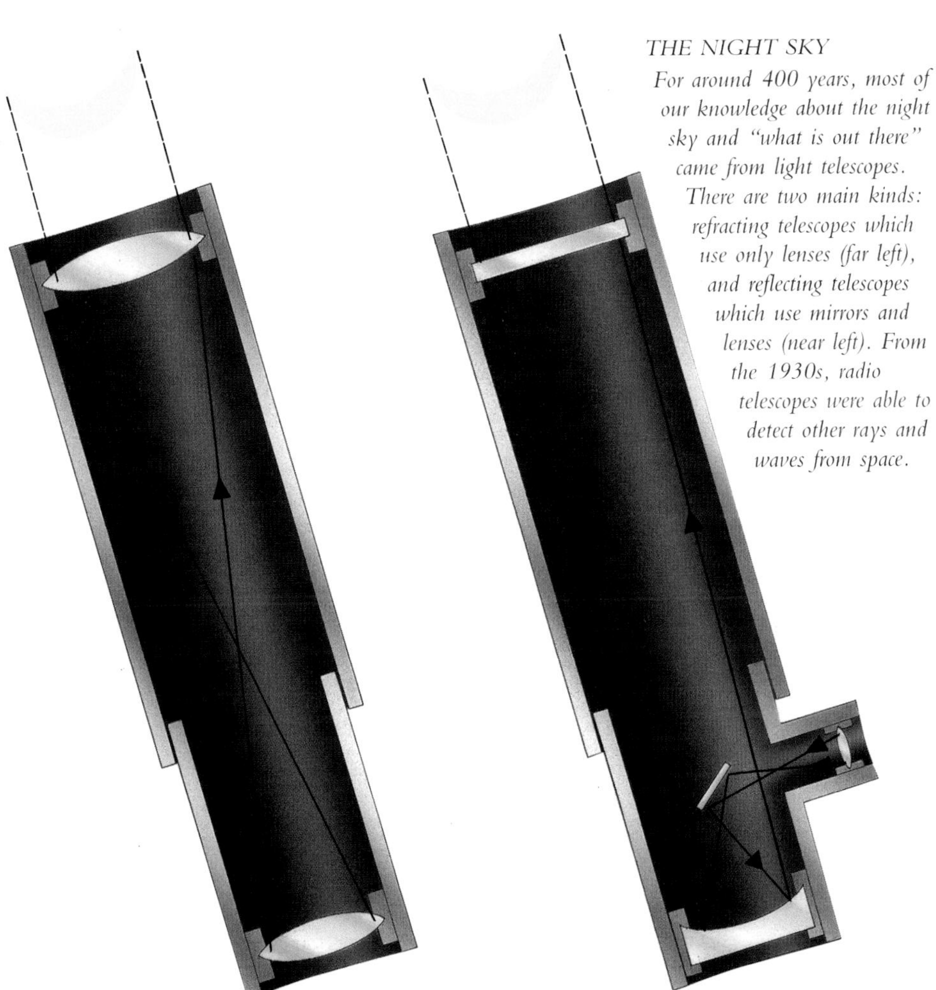

THE NIGHT SKY

For around 400 years, most of our knowledge about the night sky and "what is out there" came from light telescopes. There are two main kinds: refracting telescopes which use only lenses (far left), and reflecting telescopes which use mirrors and lenses (near left). From the 1930s, radio telescopes were able to detect other rays and waves from space.

LOOKING AND REACHING

Our quest for knowledge about space has enabled us to go beyond simply looking into space through telescopes and other instruments. The age of exploring and investigating space took a giant leap forward when the first artificial satellite, *Sputnik 1*, circled the Earth in October 1957. This was truly the beginning of the space age — the greatest adventure that the human race had ever embarked upon.

The most traveled spacecraft is *Pioneer 10*. Launched in 1972, it has reached a distance of 6.54 billion miles (10.55 billion kms). Yet that is only about 20 light-hours away. Spacecrafts have now explored all the planets in our Solar System, except the most distant planet, Pluto. Humans have so far only set foot on Earth's nearest neighbor, the Moon, which is a mere 238,330 miles (384,400 kms) away.

FIRST STEPS ON THE MOON
US astronaut Buzz Aldrin features in this famous photograph from the Apollo 11 *mission in July 1969. It was taken by Neil Armstrong, the first person to step onto the Moon's surface.*

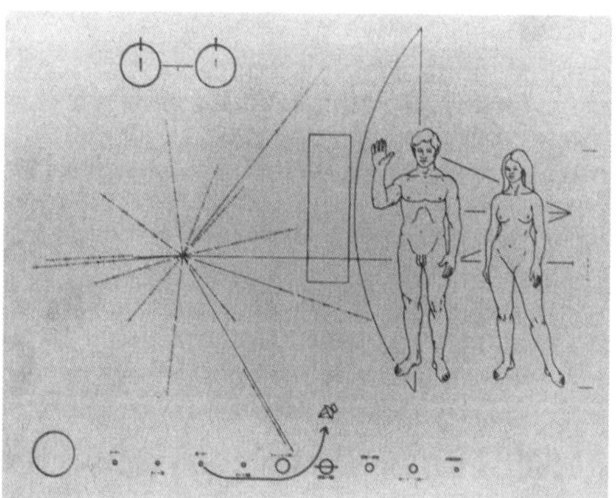

MESSAGE FROM EARTH
Pioneer 10 *carries a message plaque depicting a man and a woman, and indicating the position in space of their planet, the Earth. The plaque was placed on board in case the spacecraft encountered intelligent beings elsewhere in the Universe.*

FARTHER THAN EVER

Scientists believe that the Universe is expanding all the time. Groups of galaxies are rushing away from our Milky Way Galaxy, and also from each other. As a result, the distances between galaxies are increasing, and the Universe is getting bigger and bigger. The stars within these galaxies are slowly changing too, and new ones are constantly forming. The huge columns of cool interstellar gas and dust known as nebulae are the birthplaces of new stars. The Eagle Nebula, also known as M16, is in the constellation Serpens in the northern sky. Its tallest pillar is one light-year long, and our whole Solar System could be swallowed up inside one of the "fingertips" at the top of the column.

PLANETS AND MOONS
A montage of images taken by NASA's Voyager 1 *and* 2 *spacecraft shows the beautiful ringed planet Saturn and a few of its many moons.*

DISTANT NEBULA
This awe-inspiring image from the Hubble Space Telescope shows the Eagle Nebula, which is 7,000 light-years away from the Earth.

THE UNIVERSE

The Universe is everything that exists. It stretches farther almost than the human mind can imagine — we already know that the Universe reaches at over 13 billion light-years in every direction. The Universe is filled with matter in many different shapes, sizes, and forms. It contains dust and gases; planets such as the Earth and Jupiter; the Sun and billions of other stars; our Milky Way Galaxy and countless galaxies.

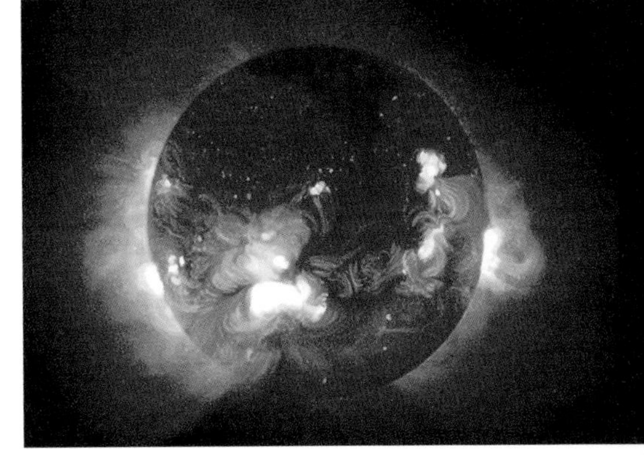

Our tiny planet Earth has a unique place in the Universe because it is the only place where we know that life definitely exists. There is still so much for us to discover about the vast Universe — about how it began and where it might end. One day we might even discover life in other parts of our Universe, or even the existence of other universes.

Home in the Cosmos

We can begin to understand the size of the Universe, and our place in it, by writing an address such as this: Jane Smith, Human Being, 5 Robins Avenue, Libertyville, North Dakota, United States, North America, The Earth, The Solar System, The Milky Way, Galaxy Group C7, The Universe. The Earth is one of nine planets that move round the Sun, forming what we call the Solar System. The Sun is just one of more than 100 billion stars in a galaxy that we call the Milky Way. The Milky Way is just one of millions of other galaxies in the vast Universe.

The Earth is a unique part of the Universe since it is the only place where we know that life definitely exists. This life is not just small microbes but a life of incredible variety. About 5 billion human beings live on Earth, yet we are just one kind of being among more than 1 million different species.

THE BEGINNING OF TIME
The powerful Hubble Space Telescope is able to look back into the "beginning of time." This image reveals a small number of the countless multicolored galaxies of all shapes and sizes that are found in the Universe.

HORSEHEAD NEBULA
The Horsehead Nebula is a cloud of cool dust. Here it is seen rising up against a backdrop of hot gas. The gas is glowing with energy from nearby stars.

THE VAST UNIVERSE

The image (above left) from the Hubble Space Telescope shows a part of our night sky. It covers an area that is just 1/30th the size of the Moon as we see it in the sky. Some of the galaxies are so far away that they are up to four billion times fainter than the limits of human vision. Although this image covers a very small area of the sky, it shows a typical arrangement of galaxies in space. In fact, from a statistical point of view the Universe looks the same in every direction.

SUN AND SOLAR SYSTEM

The Sun is a very ordinary star among billions of other stars in the Milky Way Galaxy. It has a planetary system, called the Solar System, which is made up of planets and other "left–over" material that did not form into planets. The nine known planets start with Mercury, which is at an average distance of 35.9 million miles (57.9 million km) from the Sun. Then comes Venus, the Earth, Mars, Jupiter, Saturn, Uranus, Neptune, and Pluto. Between the orbits of Mars and Jupiter lies an area of one kind of leftover material, the asteroids. These are lumps of

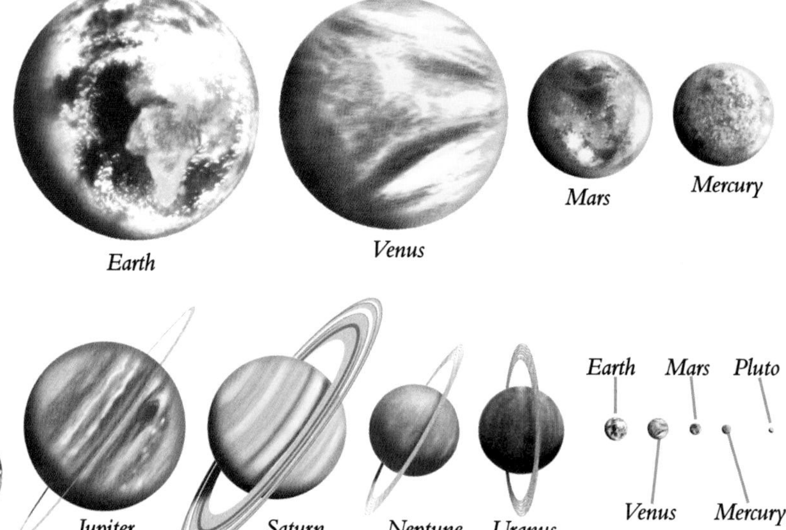

THE PLANETS OF THE SOLAR SYSTEM
This montage shows all the planets of the Solar System, compared in size. Mercury, the planet nearest to the Sun, is on the left, and Earth is third from the left. Jupiter is the largest planet. Pluto, the ninth and last planet, is almost too small to see.

rock, some of which are many miles long. Beyond Pluto is another kind of leftover material, the comets. These are bodies of rock, ice, and dust. When some comets pass close to the Sun they heat up and give off material, forming a tail. Comets also give off small rocky particles that form many of the meteors, or "shooting stars," that we see.

THE PLANETS AT A GLANCE

Planet	Average distance from Sun	Diameter
Mercury	35.9 million miles	3,026 miles
Venus	67 million miles	7,500 miles
Earth	92.8 million miles	7,909 miles
Mars	141.3 million miles	4,146 miles
Jupiter	482.5 million miles	88,536 miles
Saturn	884.7 million miles	73,966 miles
Uranus	1.78 billion miles	32,116 miles
Neptune	2.79 billion miles	30,690 miles
Pluto	3.66 billion miles	1,550 miles

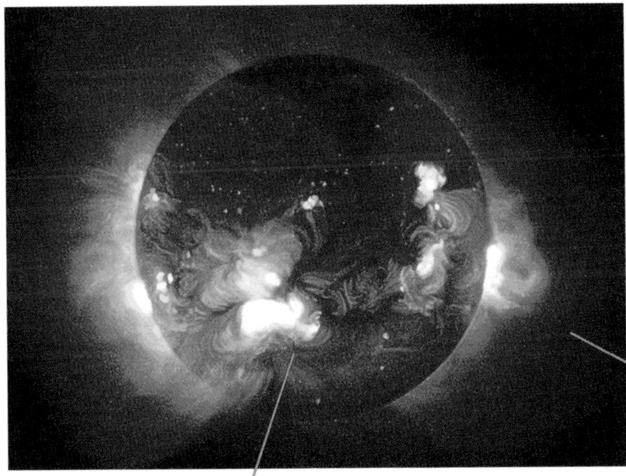

SUPER-HOT SUN
If we had X-ray eyes, this is how the Sun would look to us. This image shows the Sun's turbulent atmosphere. In the outer layer, or corona, temperatures reach as high as 2.6 million degrees Fahrenheit (2 million degrees Celsius).

The corona is the outer layer of the Sun's atmosphere

The photosphere, or surface, of the Sun

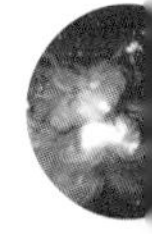

THE SPEED OF LIGHT

As we know, our Earth is a very small but rather important part of the vast Universe. The Universe is so huge that we cannot measure it in miles or kilometers, or even millions of them. Imagine writing down in miles or kilometers the distance of the Andromeda Galaxy from Earth. It would take millions of numbers — far too many to fit on a page of this book. So astronomers measure the size of the Universe or the distance of stars

THE MILKY WAY
These are some of the stars in our Milky Way Galaxy. The brightest star in our night sky is also one of the nearest. It is called Sirius, in the constellation Canis Major, and is 8.6 light-years away from Earth.

by using the speed at which light from these objects travels. Light travels in empty space at a speed of 185,871 miles (299,792 km) per second. (It travels a little slower when passing through water, glass or other denser media.) The distance that light travels in one year — 5.88 trillion miles (9.46 trillion km) — is called a light-year. Light from the Sun, which is about 93 million miles (150 million km) from Earth, takes 8 minutes to reach us. The nearest star, Proxima Centauri, is 4.225 light-years from Earth.

LIGHT ON THE MOVE
Light from the most distant galaxy so far detected in the Universe took 13.1 trillion years to reach the Earth. Light from the Sun takes just 8 minutes!

A LIGHT-YEAR

A light-year is a measure of distance, not time. It is written as 5,831,571,200,000,000,000 miles (9,500,000,000,000 km). The Moon is 238,328 miles (384,400 km) from the Earth. It took astronauts about 3 days to travel to the Moon. How long would it have taken those astronauts to travel to the brightest star Sirius? If Sirius is 8.6 light-years away, that's the same as 503,371,500,000,000,000,000 miles (811,899,530,000,000,000,000 km) After 63,363,030,000,000,000 days, or 173,597,340,000,000 years, they would reach Sirius!

THE ANDROMEDA GALAXY
The nearest galaxy to our Milky Way Galaxy is the spectacular Andromeda Galaxy. It can just be seen with the naked eye as a small fuzzy patch of light in the northern skies, near the constellation Pegasus.

THE MILKY WAY

THE MILKY WAY GALAXY
No one knows exactly what the Milky Way Galaxy looks like from the outside. It is a spiral-shaped galaxy, with arms that resemble the shape of a spinning firework.

We can see about 5,000 individual stars in the night sky. On a really clear night, it is also possible to see one or two galaxies. The most visible part of our own Milky Way Galaxy is the part of the sky that looks like a misty cloud. It is really a band of millions of stars. This part of our Galaxy that we can actually see is also called the Milky Way. When we look at it, we are looking towards the center of our Galaxy where there are the most stars. We cannot actually see the center because we are situated a long way from the center in the Orion arm, an outer arm of the Galaxy. When we look at the region where the famous Orion constellation is situated, there seem to be fewer stars. This is because we are looking toward the edge of the Galaxy.

HALLEY'S COMET
This photo of the Milky Way also shows Halley's Comet passing across the sky. A comet orbits the Sun, and so it is much closer to Earth than the distant stars.

OUR GALAXY

If we could look at it from the side, the Milky Way Galaxy would be a rather thin disk with a dense, bulging center surrounded by a halo. This halo is a mass of older stars surrounding the central mass, which contains the most stars. The thinner edges of the disk are the outer arms of the Galaxy. The Milky Way Galaxy is about 2,000 light-years thick but 100,000 light-years across. It contains an estimated 100,000 million stars. Our Sun, which is one of those stars, is about 30,000 light-years from the center. It orbits the center of the Galaxy at a speed of 170 miles (274 km) per second.

BIRTH OF THE STARS
In addition to the millions of stars in the Milky Way Galaxy, there are also many nebulae, which are regions where stars are being born. This is a nebula in the constellation Cygnus.

GALAXIES AND NEBULAE

When we look into the night sky, we can observe other objects in addition to stars. Some, like nebulae, are inside the Milky Way, while others are distant galaxies much farther away. The most famous nebula is part of the constellation Orion (the Hunter). Hanging from the hunter's "belt" of three stars is a "sword" of stars. In the middle of the sword is the Orion Nebula, a region of hot gas and dust where new stars are born. The nebula is 1,500 light-years away and 15 light-years across.

The most famous galaxy is the Andromeda Galaxy, which is close to the constellation Pegasus. It is the most distant object visible to the naked eye and looks like a small fuzzy patch. The Andromeda Galaxy is a huge spiral galaxy, 2.2 billion light-years away. It is thought to contain at least 300 billion stars.

The Egg Nebula is a swirling cloud of gas and dust

SOMBRERO GALAXY
This is a spectacular, almost perfectly formed galaxy called the Sombrero Galaxy. It is also called a nebula because it has a prominent central band of dust and cloud.

EGG NEBULA
This strange-looking nebula, called the Egg Nebula, is 3,000 light-years away from Earth. This photo was taken by the Hubble Space Telescope, which has helped to revolutionize astronomy.

DYING STARS

Some nebulae are the remnants of dying stars. The mysterious "searchlight" beams emerging from a hidden star in the Egg Nebula (left) are being crisscrossed by many bright arcs. The nebula is a huge cloud of dust and gas ejected by a dying star. The star, which is hidden by a dense cocoon of dust, is expanding at a speed of 12 miles (20 km) per second. Eventually the dying star will blow off its outer layers to form a planetary nebula.

THE VAST UNIVERSE

Scientists estimate that there are billions of galaxies in the Universe, but we cannot be sure because no one knows where the Universe ends — if it does! People who believe that the Universe was created by a god or gods can accept that it is beyond our human understanding. There are also many people who believe that the Universe was not created but just "happened," perhaps in a "big bang"; its glory was simply an accident. Whatever people think, the human race will perhaps never fully understand the nature of the Universe. However, this lack of understanding will not stop us from

SPACE AND TIME
Space is not straight, continuous, or constant. Neither is time. In the relationship between space and time, space curves around massive objects, such as stars, while time speeds up or slows down. The Universe is filled with mysterious objects such as black holes, and perhaps even "short-cuts," or wormholes, to other dimensions or even other universes.

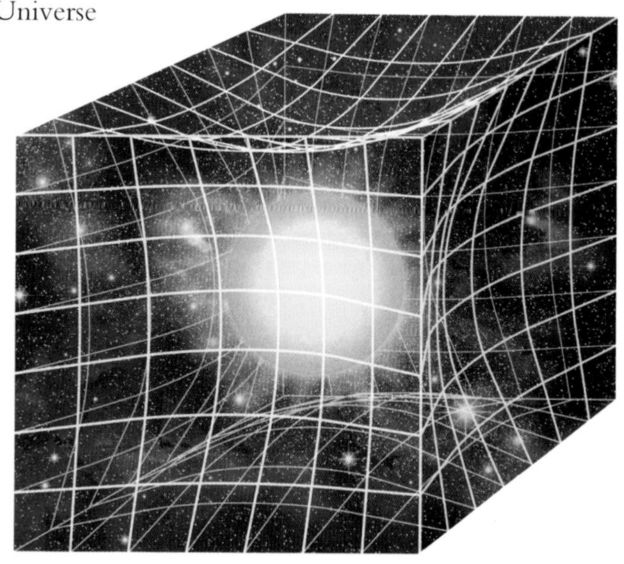

trying to find out more about the Universe, through observation and exploration. Yet each new and exciting discovery seems to create even more questions to be answered!

BIGGER AND FARTHER

In 1990, astronomers discovered a galaxy called Abell 2029 in the constellation Virgo. The Abell 2029 Galaxy is 1,070 million light-years away and has an incredible diameter of 5.6 million light-years, which is 80 times bigger than the Milky Way. The most remote object yet discovered in the Universe is over 13 million light-years away and is a quasar inside a galaxy. Quasars are mysterious space objects which are thought to be associated with equally mysterious black holes.

THE VAST UNIVERSE
We may only be able to see a very small section of a much larger Universe. The size of the Universe is so great that human beings can only begin to perceive it.

EVERYTHING THAT EXISTS
The Universe contains everything that exists. It contains all of space, time and matter. There could be millions of other galaxies like our Milky Way in the Universe. The most distant galaxy that we have detected so far is over 13 million light-years away.

SIGNPOSTS IN THE SKY

Long ago, people divided up the night sky into distinct areas called constellations. The word "constellation" also refers to a pattern of stars that appears in a particular area of the sky. Many of these star patterns are named after animals or figures from mythology.

Astronomers have named a total of 88 different constellations. The most famous one in the night sky of the northern hemisphere is probably Ursa Major, or the Great Bear. It contains a well-known group of stars called the Big Dipper. Even people who are not interested in astronomy can usually locate the Big Dipper easily. It is sometimes described as being like a saucepan with a bent handle. It has four main stars which form an outline rather like that of a saucepan. The bent handle consists of another three main stars.

THE BIG DIPPER
The Big Dipper makes a wonderful signpost in the sky. The two stars on the right-hand side of the saucepan point up to a star called Polaris. This is the Pole Star, the nearest star to the place in the northern skies to which the Earth's axis points. Because the Earth rotates on its own axis, the Pole Star looks as if it remains in the same place in the sky as the other stars move around it.

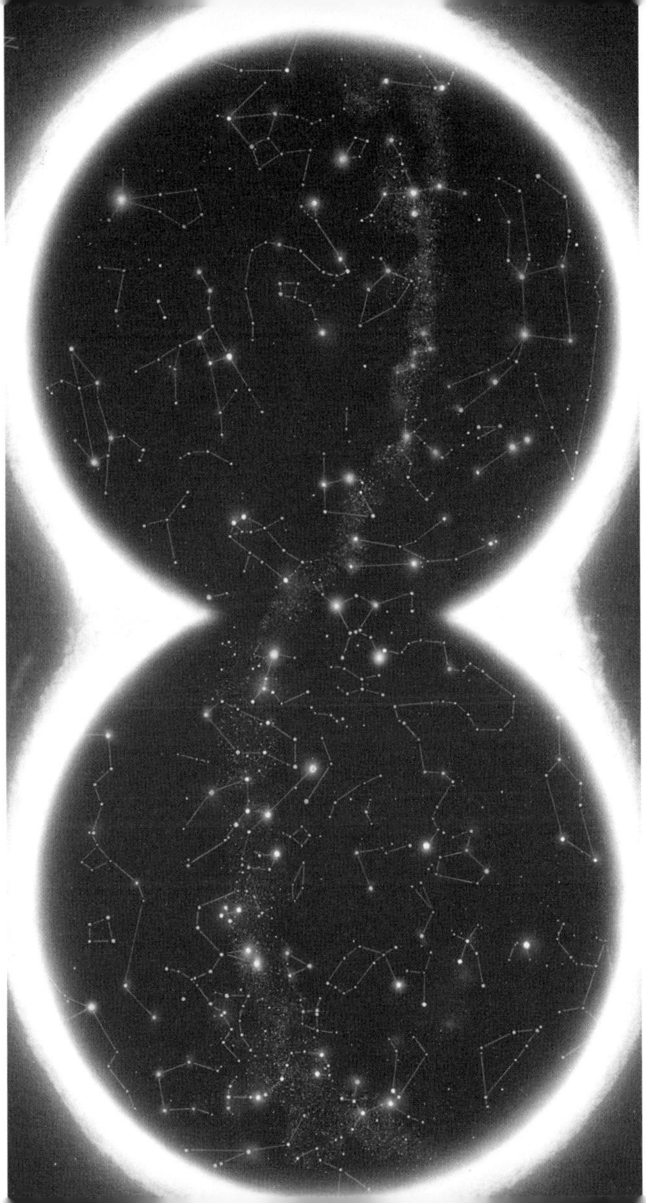

ORION AND PEGASUS

The magnificent constellation Orion dominates the winter sky in the northern hemisphere. Not only can it be used as a signpost, but it is also one of the constellations that really seems to look like the character it depicts—the Hunter. We can clearly make out his belt and sword, and to the right his bent bow and arrow pointing at the horns of an angry bull—Taurus. The Orion Nebula is visible within the third star of the Hunter's sword. The famous Andromeda Galaxy can be located using the almost perfect "square" set of stars in the constellation Pegasus.

THE NIGHT SKIES
This map shows the stars and constellations of the northern (top) and southern (bottom) hemispheres. The sky in the northern hemisphere is shown as it appears from the North Pole. Polaris, the Pole Star, is directly overhead. There is no equivalent of the Pole Star in the southern hemisphere.

BRIGHTEST STARS

The measure of a star's brightness is called its magnitude. The smaller the magnitude, the brighter the star is. The first person to work out a level of brightness was probably a Greek astronomer called Hipparchus. He divided the stars as he could see them into six groups. The stars in the brightest group were first magnitude, and the faintest stars were sixth magnitude. Later, other astronomers worked out that Hipparchus's brightest stars were about 100 times brighter than sixth-magnitude stars. The stars that were 100 times brighter than sixth magnitude were given a minus number, and the fainter ones were given a plus number. One of the brightest stars, Arcturus, the main star of the constellation Boötes, has a magnitude of -0.06.

STAR BRIGHT

This dramatic picture shows a high-resolution image of a star. Out of the 14 brightest stars in the night sky, 11 of them are visible in the northern hemisphere.

WITH THE NAKED EYE

We can see about 5,000 stars in the night sky with the naked eye. Fainter stars can be seen only with binoculars or telescopes.

THE 14 BRIGHTEST STARS		
Star	**Constellation**	**Magnitude**
Sirius	Canis Major	–1.45
Canopus	Carina	–0.73
Alpha Centauri	Centaurus	–0.1
Arcturus	Boötes	–0.06
Vega	Lyra	–0.04
Capella	Auriga	–0.08
Rigel	Orion	–0.11
Procyon	Canis Minor	+0.35
Achernar	Eridanus	+0.48
Beta Centauri	Centaurus	+0.6
Altair	Aquila	+0.77
Betelgeuse	Orion	+0.8
Aldebaran	Taurus	+0.85
Acrux	Crux	+0.09

BETELGEUSE

Betelgeuse, in the constellation Orion, is an old and dying red giant star. This picture was taken by the Hubble Space Telescope. Betelgeuse is about 310 light-years away from Earth. It has a diameter of about 300 million miles (500 million km) and could swallow up the Solar System almost as far as the planet Jupiter.

MOVING STARS

The Earth's spinning motion makes the stars seem to move across the night sky. The exception is the Pole Star, which seems to stay in the same place with the other stars revolving around it. The southern night sky has no Pole Star, and the area in the sky to which the Earth's South Pole points is almost empty of bright stars. The long axis of the famous Southern Cross (Crux) constellation in the southern sky points to the south celestial pole, the sky's south pole.

You are probably used to seeing a constellation, such as Orion, in the northern night sky. If you move closer to the Equator you will see the constellation on its side. As you move south, it will appear upside down and eventually disappear the farther south you travel. The same phenomena will happen to the southern stars as you travel north. The farther south you travel, the more of the southern sky you will see and the less of the north.

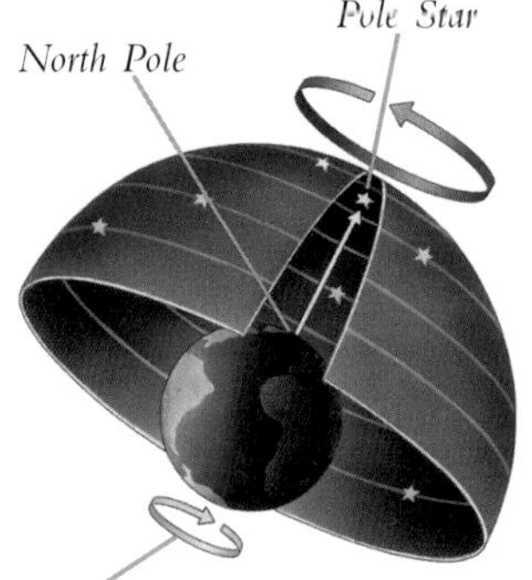

AT THE NORTH POLE
If you stood on the North Pole and looked up at the night sky, the Pole Star would be directly above your head. Because the Earth's axis points almost directly to the Pole Star, the other stars seem to revolve around it as the Earth rotates.

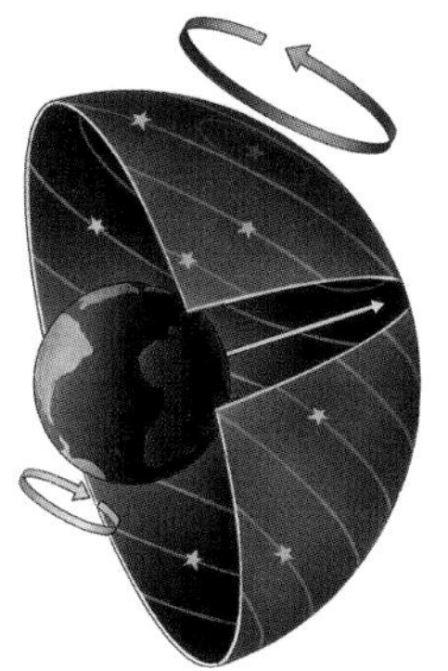

NORTH OF THE EQUATOR
If you stand on the Earth somewhere between 10 and 20 degrees north of the Equator, you will be in the mid-northern latitudes. The stars overhead will still seem to be moving, slowly changing the appearance of the sky from night to night.

MAGELLAN'S CLOUDS

The Large and Small Magellanic Clouds are two small galaxies visible in the southern hemisphere. They are named for a Portuguese navigator called Ferdinand Magellan, who described them on his travels in 1521. Both galaxies are about 160,000 light-years away, which is about 12 times closer than the Andromeda Galaxy. The Large Magellanic Cloud contains the largest known nebula, 30 Doradus, which is also called the Tarantula Nebula.

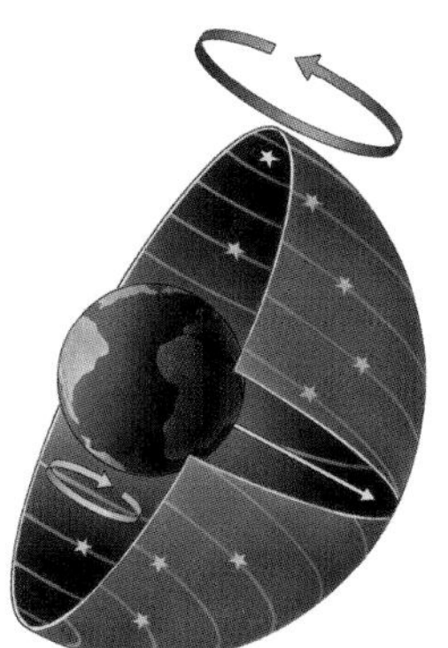

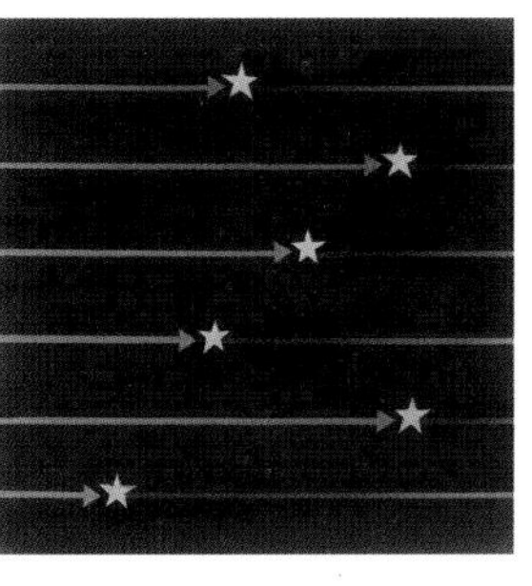

AT THE EQUATOR
An observer who stands at the Equator will see all the stars in the night sky during the period of a year. At the Equator, the stars seem to move in straight lines across the sky.

EXPLORING THE UNIVERSE

Long before telescopes were invented some 400 years ago, humans were fascinated by what they saw in the night sky. Early astronomers could only guess at what lies beyond our Earth. Modern astronomers are armed with sophisticated equipment that allows them to see deeper into the Universe than ever before.

Astronomy has made enormous advances since the time when

Galileo made his first observations using a telescope in the early seventeenth century. Giant observatories on the ground and satellites in orbit above the Earth enable today's professional astronomers to discover new galaxies and to watch the dying stages of a distant star. With new and increasingly powerful telescopes, they can observe mysterious phenomena such as black holes and quasars.

AGES OF ASTRONOMY

Ancient peoples could only use their own eyes to gaze in wonder at the night sky. They named constellations and some stars, and measured the movements of the Moon and the Sun. Later, telescopes were invented and astronomers became aware that the Earth was not the center of the Universe but was one small planet traveling around a small star.

New telescopes have now been developed that observe the Universe not as we would see it with our eyes but using different wavelengths, such as X-rays and ultraviolet light. Some of these telescopes have been launched into space to obtain a better view of the Universe, without the interference of the Earth's atmosphere. Astronomers believe that the Universe may have been formed in a "Big Bang", and that it is still expanding. They are now searching farther than ever into the deepest parts of the Universe.

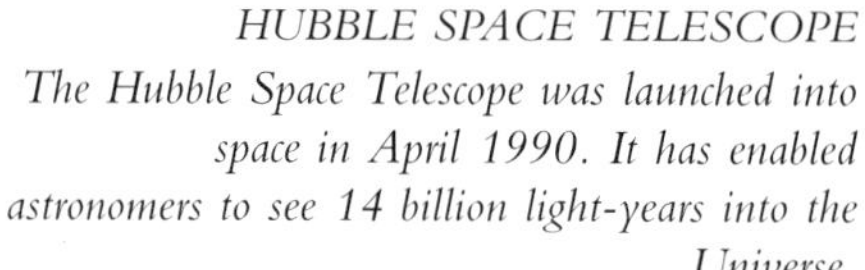

HUBBLE SPACE TELESCOPE
The Hubble Space Telescope was launched into space in April 1990. It has enabled astronomers to see 14 billion light-years into the Universe.

BACK TO THE BEGINNING
In 1609, the astronomer Galileo saw mountains on the Moon, the phases of Venus, and four of the moons of Jupiter. Almost 400 years later, we are able to look halfway back to the beginning of the creation of the Universe.

NEW IDEAS

Galileo was punished by the Catholic Church for his revolutionary ideas on astronomy. He realized, like some other astronomers, that the old idea of the Earth being at the center of the Universe were wrong. He was brought before an Inquisition and forced to deny his newly found beliefs. Galileo lived quietly for the rest of his life near Florence but he continued with his research. His last book was smuggled out of Italy and published in the Netherlands, in 1638. The book discussed physical mechanics and the principles of physics.

GALILEO GALILEI (1564–1642)
The Italian mathematician, astronomer, and physicist Galileo Galilei was one of the first people to study the night sky with a telescope, in 1609.

THE FIRST ASTRONOMERS

Early astronomers mapped the night sky, named constellations, and observed the movements of the Sun and Moon. They built temples, simple observatories, and other buildings to carry out their work. The Egyptian pyramids were built so that they were aligned with certain parts of the sky. The first calendars were being created by about 2400 B.C. The Greeks named the constellations after figures from mythology. For example, Orion the Hunter is attacking Taurus the Bull, who has captured the seven sisters, the Pleiades star cluster. Some of the constellations were linked to the Zodiac, the path across the sky along which the Sun travels, and this gave rise to astrology.

Gradually, through observations and measurements, astronomers such as the Greek Ptolemy reached a better but not necessarily correct understanding about the Earth's place in space.

STONE CIRCLES

Stone circles such as Stonehenge in England were aligned to mark time using the movement of the Sun. Stonehenge was built in about 1800 B.C., to match the rising and setting of the Sun at the summer solstice. The three separate circles were all linked to astronomical observations and could be used to calculate calendar dates.

MEASURING THE STARS
Early astronomers used an instrument called an astrolabe to measure the height of stars above the horizon. An astrolabe consists of a metal disk inside a round frame, with a sighting arm across the middle. A scale is marked around the outside of the frame.

USING ASTRONOMY

- The first use of astronomy was to measure time.
- The early Egyptians used the appearance of the brightest star, Sirius, to mark the start of the season when the Nile River was in flood.
- The Greek astronomer Ptolemy made accurate measurements of the movements of the planets in about A.D. 180. He concluded that the Earth lay at the center of the Universe.
- Some Native American tribes laid out rings of stones to mark the movements of the Sun and the stars.

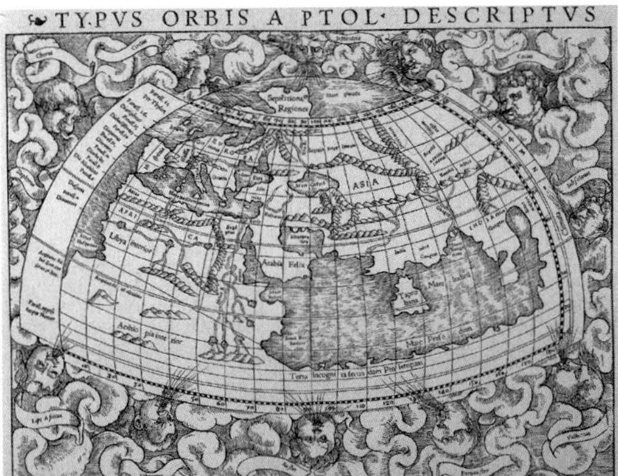

THE EARTH-CENTERED UNIVERSE
Astronomers in the Middle Ages believed that the Earth was at the center of the Universe. They thought that the planets moved around the Earth.

RENAISSANCE ASTRONOMY

Galileo is often credited with the development of the first telescope, but he actually designed his instrument based on the drawings of others. He was, however, the first to make workable observations using a telescope that magnified objects by up to 30 times. During his work Galileo observed that Venus had phases and that four moons traveled around Jupiter. He then realized that the idea of everything, including the Sun, moving round the Earth was wrong. Ptolemy's theory, which had been accepted for hundreds of years, was not true.

JOHANNES KEPLER (1571–1630)
German astronomer and mathematician Johannes Kepler discovered how the planets travel around the Sun. He worked out that they move round in oval-shaped paths called orbits.

NIKOLAI COPERNICUS (1473–1543)
The Polish astronomer Nikolai Copernicus was the first astronomer to understand that the Earth and the other planets move around the Sun.

ISAAC NEWTON (1642–1727)
Englishman Isaac Newton discovered the law of gravitation at the age of 22. He realized that gravity is the reason why the planets, for example, orbit the Sun.

The Earth was not the center of the Universe. Copernicus proved mathematically that the Earth was a planet orbiting the Sun.

Isaac Newton discovered how orbits were made possible and published a work on the universal law of gravitation almost 20 years after his discovery. According to Newton, the forces that acted on planets orbiting the Sun were the inertia of their own speed and the pull of the Sun's gravity.

EDMOND HALLEY (1656–1742)
Englishman Edmond Halley is famous for his work on comets. He predicted that a comet he saw in 1682 would return in 76 years' time. He was right, and it became known as Halley's Comet.

GRAVITY RULES

It is said that Isaac Newton came to his conclusion about gravity when an apple fell on his head – or perhaps nearby – while he was sitting under a tree. He began to think about what had happened, and many months later he realized that it is not only the Sun that exerts a gravitational force on the Earth. Objects fall on the Earth as a result of the gravitational power of the Earth's mass. Newton calculated the pull of the Earth on the Moon, and explained why the Moon travels as fast as it does and why it is at a certain distance from the Earth. Newton also designed the first reflecting telescopes.

HOW TELESCOPES WORK

A basic refracting telescope uses two lenses, one at each end of the telescope. Light from an object falls on the larger glass lens, called the objective. This lens "bends", or refracts, the rays of light and brings them together at a point called the focal point. This point coincides with the position of the smaller lens, called the eyepiece. The eyepiece lens magnifies the image. The distance between the objective and the focal point is called the focal length.

A reflecting telescope uses a mirror, instead of a lens, as its objective. The mirror is concave and reflects light rays at the focal point. A smaller secondary mirror then reflects the image onto the eyepiece lens. The Newtonian reflecting telescope has a small mirror that reflects the image to an eyepiece on the side of the telescope. This kind of telescope is used by most amateur astronomers.

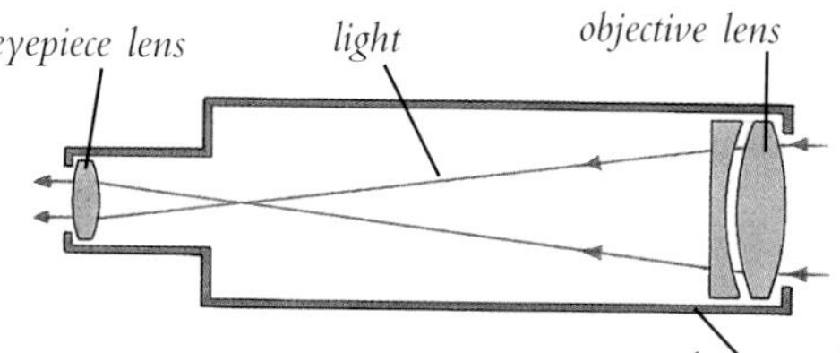

A REFRACTING TELESCOPE
Light from an object in the night sky falls on a glass lens at the front of the telescope. The light rays are focused onto a second lens that magnifies the image for the observer.

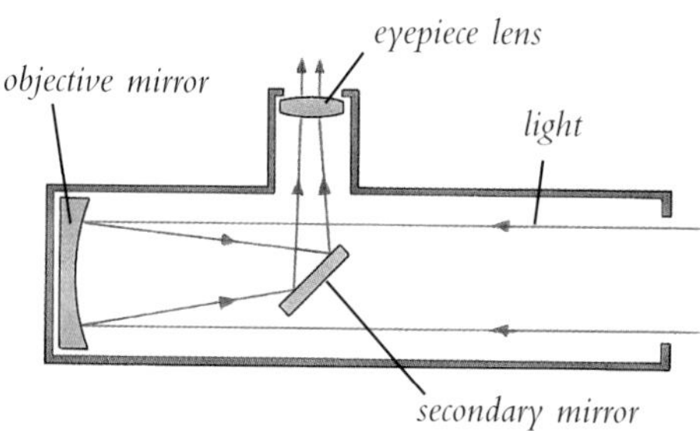

A REFLECTING TELESCOPE
Reflecting telescopes are better than refracting ones because they do not distort the light as much. Also, they can be made with wider, or larger, apertures (openings), which make them more powerful.

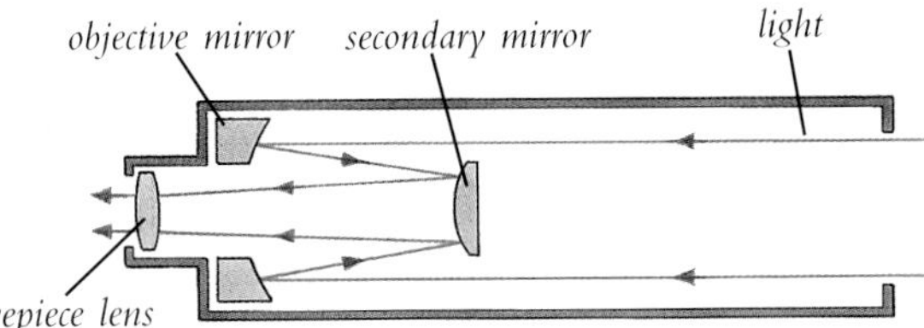

A CASSEGRAIN REFLECTING TELESCOPE
The Cassegrain telescope has a curved secondary mirror that is about one quarter of the diameter of the objective mirror. The objective mirror has a small hole in its center. The small mirror reflects light back through the hole in the objective mirror and onto the eyepiece lens.

FAMOUS NAMES

Many of the most famous astronomers would have been pleased to know that today's spacecrafts are named after them. One of the first astronomical space observatories was called *Copernicus*. The *Galileo* spacecraft is orbiting Jupiter and taking dramatic pictures, particularly of the four moons that Galileo himself saw with one of the first telescopes. The *Cassini–Huygens* spacecraft is on its way to become the first craft to orbit Saturn. It carries a piggyback craft that will attempt to land on one of the moons of the ringed planet.

Larger astronomical observatories use another type of reflecting telescope called the Cassegrain telescope. Its small curved secondary mirror reflects the light back through a hole in the center of the objective mirror and onto the eyepiece lens at the end of the telescope. This design increases the telescope's focal length, the distance between the center of the objective mirror and the focal point.

WILLIAM HERSCHEL (1732–1822)
Sir William Herschel started observing the night sky in 1771, using a small reflecting telescope. He discovered the planet Uranus, the first planet to be discovered with the help of a telescope.

BIGGER TELESCOPES

The development of the first powerful telescopes, and the observations made using them, created a revolution in astronomy. They had as dramatic an effect on our knowledge of the Universe as the images from the Hubble Space Telescope have on us today. After Galileo, many famous astronomers, such as Christiaan Huygens and Giovanni Cassini, started to make rather basic observations. Huygens made a drawing of Mars in 1659, while Cassini observed Saturn's rings in 1676. The Earl of Rosse observed galaxies and nebulas with his large telescope, and named the Crab Nebula in 1848. By the twentieth century, huge observatories were being built, including the Palomar Observatory in California. This became the world's most powerful observatory, and was able to take dramatic images of objects such as the Orion Nebula.

KITT PEAK NATIONAL OBSERVATORY
The McMath solar telescope at the Kitt Peak National Observatory in Tucson, Arizona, USA has a 60-inch (159 cm) main mirror with a focal length of 300 feet (91 m).

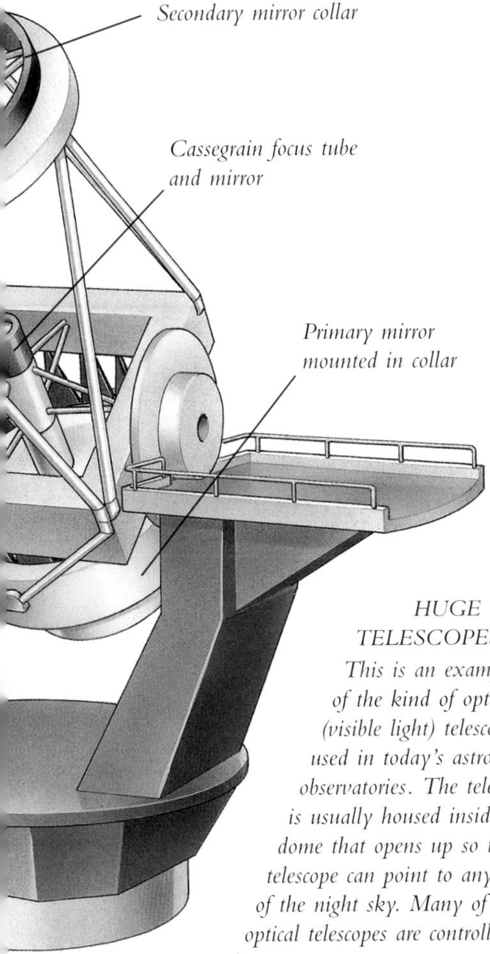

Secondary mirror collar

Cassegrain focus tube and mirror

Primary mirror mounted in collar

HUGE TELESCOPES
This is an example of the kind of optical (visible light) telescope used in today's astronomical observatories. The telescope is usually housed inside a giant dome that opens up so that the telescope can point to any part of the night sky. Many of these optical telescopes are controlled by computers.

BIGGER THAN BEFORE

• Irish astronomer William Parsons, the Third Earl of Rosse, built the famous 70-inch (183-cm) diameter reflecting telescope in 1845. It was the largest of its kind in the world until 1917.

• The 41-inch (102-cm) diameter Yerkes refractor telescope at Williams Bay, Wisconsin, is still the largest of its kind today. It was the brainchild of George Ellery Hale, an enthusiastic astronomer who had developed small observatories with the help of his father. In 1892, Hale was assisted by Charles T. Yerkes who provided $34,900 to build the world's largest refracting telescope. The telescope was completed in 1897.

• The Multiple Mirror Telescope in Arizona, has six separate mirrors. A computer adjusts the mirrors so that they focus all the light at a single point.

• The Hale Telescope, one of the largest reflecting telescopes to be built, is at the Palomar Observatory in California. Built in 1945, it has a 200-inch (508-cm) diameter reflecting mirror and is mounted in a dome that can rotate.

• The world's largest optical telescope is the Keck telescope, Hawaii, which has a 33 foot (10-m) wide mirror. It is set in a huge dome on a mountaintop in Hawaii.

Other Rays and Waves

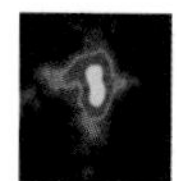

The light that we see from the Sun is only part of the huge amount of energy, or radiation, given off by the Sun. This radiation consists of waves of energy called electromagnetic waves. They include radio waves, microwaves, infrared rays, light rays, ultraviolet rays, X-rays, gamma rays, and cosmic rays. Using telescopes on the ground we can see visible light rays and also detect some radio waves and microwaves. The Earth's atmosphere blots out other kinds of radiation, which is a good thing since most of it is harmful to living things.

In 1931, a radio engineer named Karl Jansky, from Oklahoma, was trying to work out why "static" hisses were interfering with radio communications. He built a special antenna and picked up a continuous hiss, which he discovered came from the Milky Way. Radio astronomy was born. The noises that Jansky detected were signals from very distant sources of radio waves, such as nebulae. Because nebulae are often millions

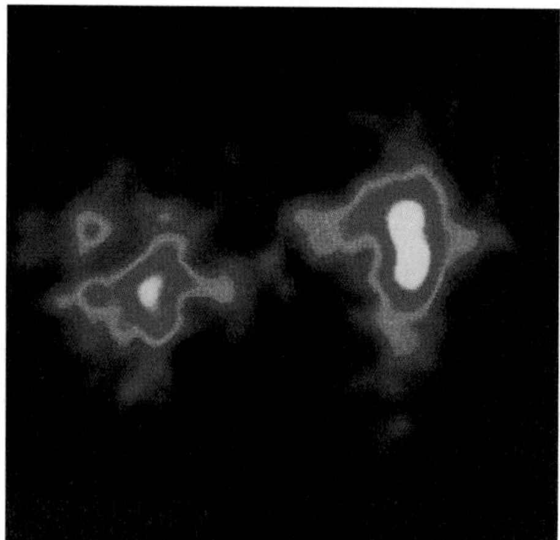

X-RAY IMAGE

This is an X-ray image of the Andromeda Galaxy. X-ray images of the Universe are particularly important. They enable astronomers to "see" extremely high-energy events, such as explosions that generate very high temperatures.

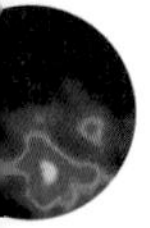

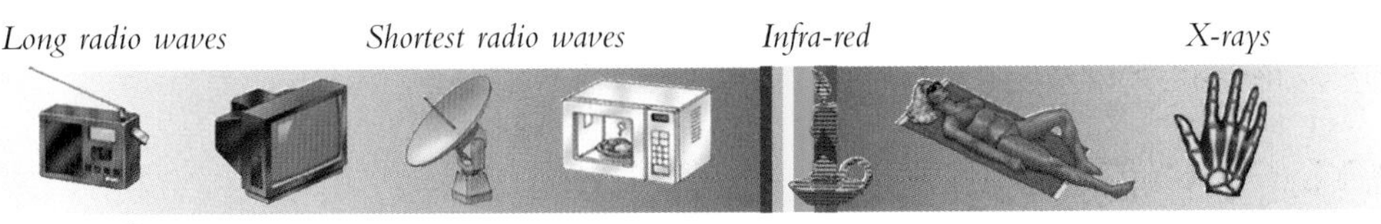

THE ELECTROMAGNETIC SPECTRUM
The visible light that we can see is one kind of electromagnetic energy that makes up the electromagnetic spectrum. Waves of electromagnetic energy have different lengths. Radio waves have a long wavelength. X-rays have a short one. Visible light energy consists of different primary colors, which we can see in a rainbow.

of light-years away, it takes a long time for the radio waves to reach the Earth. Radio astronomy is an extremely important part of astronomy. It can be carried out mainly using large telescopes built on the ground, since most radio waves can penetrate the Earth's atmosphere. These telescopes have huge dishes that focus the radio waves onto the tip of a metal rod in the center of the dish.

SPECTROGRAPHS

Using an optical telescope, it is possible to split the light coming from an object into its range, or spectrum, of different colors. This is similar to the way in which a prism will split light into the colors of a rainbow. Each color has a different wavelength — violet has the shortest and red has the longest. The astronomical instrument that splits the light in a telescope is called a spectrograph. It can be used to analyze starlight and to identify its elements. In this way a chart of categories of stars can be created, ranging from blue hot stars to red cooler stars.

VISIBLE LIGHT IMAGE
A visible light image such as this one of a galaxy looks quite beautiful to us. However, the information we receive from the image may be very little compared to what we would find out if we took an image of the same galaxy using X-ray, infrared, or ultraviolet waves.

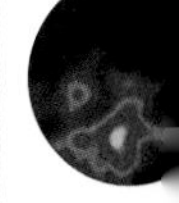

SPACE TELESCOPES

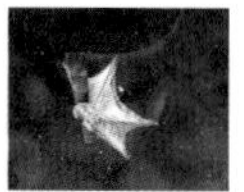

Two small "windows" in the electromagnetic spectrum allow observations of space to be made from the Earth: visible light waves and radio waves both pass through the Earth's atmosphere. The other waves of the spectrum are largely obscured by the atmosphere. Rockets carrying astronomical instruments are able to make brief flights above the atmosphere but only satellites can provide continuous flight beyond the atmosphere. The first satellite, *Sputnik 1,* was only launched in 1957, and the first astronomical satellite followed four years later. Today, astronomical satellites of almost every type have been launched, from the optical Hubble Space Telescope to the Compton Gamma Ray Observatory. Other exciting new projects which are in the pipeline

NASA Goddard

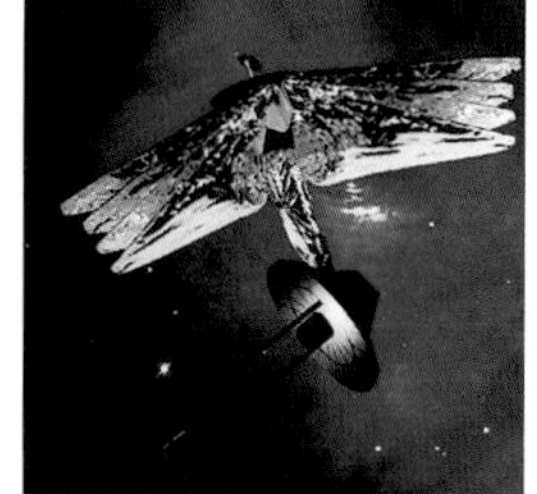

Ball Aerospace

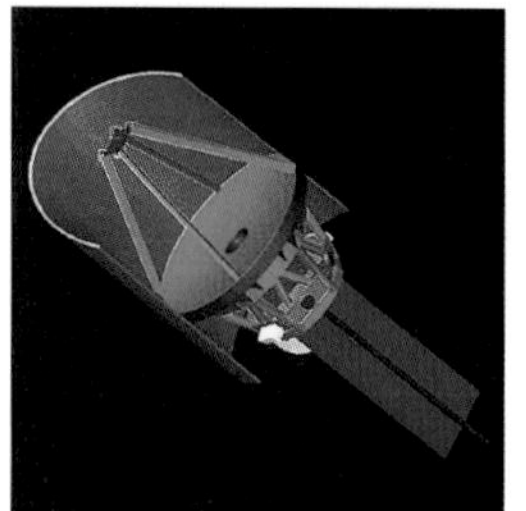

Lockheed Martin

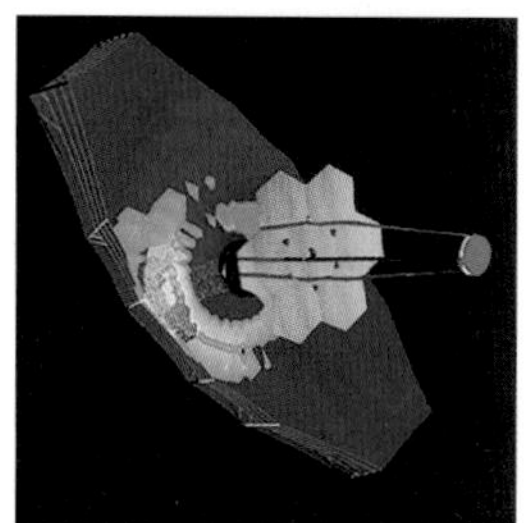

TRW

INTO THE FUTURE

The Next Generation Space Telescope will succeed the Hubble Space Telescope. It will be so powerful that it will be able to photograph planets moving around other stars. These illustrations depict four concepts of what the Next Generation Space Telescope might look like.

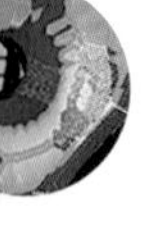

include the European X-Ray Multi-Mirror mission. Scientists still hope that an astronomical base on the Moon might be possible one day. By observing objects in all the different wavelengths of the electromagnetic spectrum we can build up a more complete picture of the same object.

EXPLORER 11

The first dedicated astronomical satellite, called *Explorer 11,* was launched on April 27, 1961. The 81-pound (37-kg) spacecraft was equipped with a gamma-ray telescope. The satellite could detect gamma-ray sources, particularly in the Milky Way. However, the *Juno 2* booster sent the satellite into a higher orbit than intended. As a result, the satellite was flying through the Earth's own radiation belts, which saturated its instruments for 60 percent of each orbit.

A SUPER-TELESCOPE
By launching several telescopes to orbit together, a larger-aperture telescope can be created. This is the concept behind a mission currently being planned by NASA.

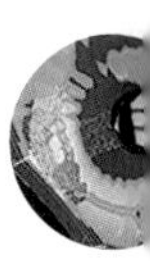

HUBBLE SPACE TELESCOPE

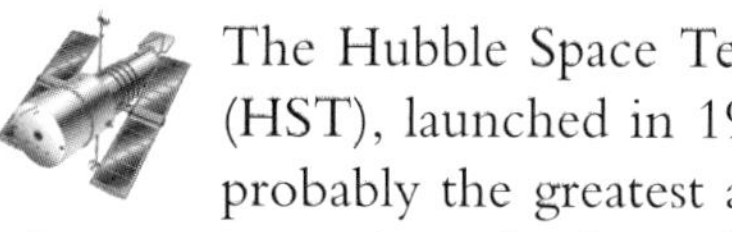

The Hubble Space Telescope (HST), launched in 1990, is probably the greatest advance in astronomy since the first telescope was used by Galileo. It can see 50 times deeper into space than the most powerful telescope on Earth. Using a telescope on the Earth to see into space is rather like being underwater in a swimming pool and trying to see outside the pool. It is very difficult to see through the thick atmosphere. In space, the HST is also not subject to the slight "bending" by gravity that affects telescopes on the Earth. The control system on the HST, which controls its pointing, is so accurate that it could identify a small coin in New York from a point in Washington.

SERVICE IN SPACE
The HST was designed to be serviced in space by astronauts from the space shuttle. There have been two servicing missions so far, and a third one is scheduled for 2000.

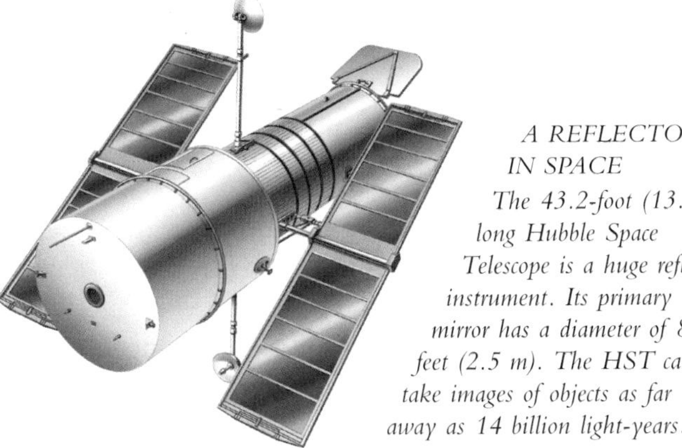

A REFLECTOR IN SPACE
The 43.2-foot (13.1-m) long Hubble Space Telescope is a huge reflector instrument. Its primary mirror has a diameter of 8.2 feet (2.5 m). The HST can take images of objects as far away as 14 billion light-years.

SPECTACLES FOR HST

When the vital primary mirror of the HST was manufactured in the 1980s, a very slight mistake was made in its precise shaping. As a result, when the HST was first used in orbit, its images were a little blurred and were not quite the quality that had been planned. An ingenious repair was made by Space Shuttle astronauts in 1993. It was rather like placing a pair of eyeglasses over the mirror to restore it to its planned quality. The primary mirror's surface is so smooth that if it were the size of the whole of the United States, any very slight bumps on it would be no bigger than a saucer.

With the help of the HST, astronomers are now able to see sights they used to dream of — black holes, quasars, and even possible planets moving around other stars. Many of the visible light images are combined with other observations made in other wavelengths by the HST, producing more complete photos.

BUILDING HUBBLE
Although the Hubble Space Telescope was built in the USA, many international companies supplied important equipment for it. The United Kingdom supplied the first and second set of solar arrays to provide the telescope with electricity.

VIEWS FROM HUBBLE 1

When the first image of a galaxy taken by the Hubble Space Telescope after its space repair was released, it excited astronomers all over the world. The HST images revealed extraordinary and beautiful views of deep space — images which in some cases are beyond our full understanding. Images of the aftermath of the explosive death of some stars, and of the process of stellar death, have revealed nebulas of the most unique and intricate shapes.

SUPERNOVA BLAST WAVE
The Cygnus Loop marks the edge of a bubble-like, expanding blast wave from a huge stellar explosion. It occurred about 15,000 years ago.

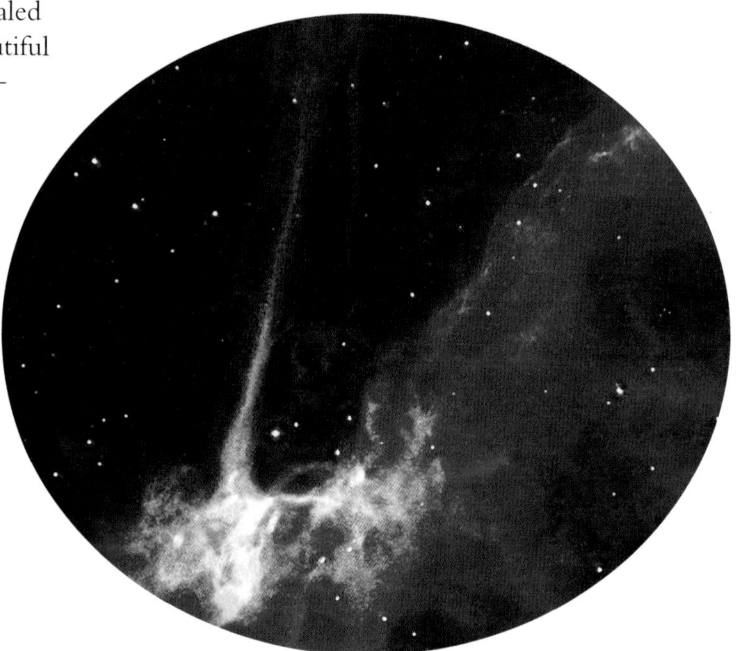

BEFORE AND AFTER
These pictures show the M100 galaxy taken by the Hubble Space Telescope shortly after it was launched in 1990 (left), and in 1994 (right) after the primary mirror was corrected by Shuttle astronauts.

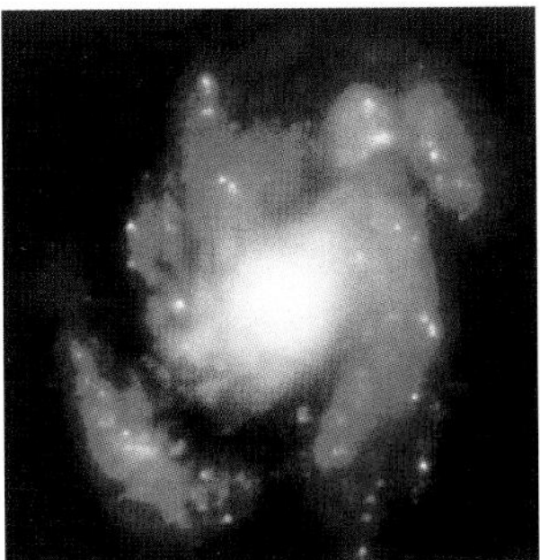

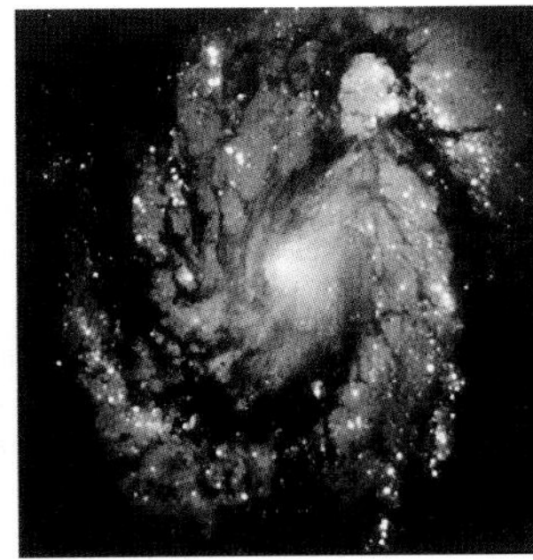

Astronomers still cannot explain these images fully. Their beauty perhaps conceals the enormously powerful forces at work, with gases being expelled at speeds of several thousand miles per second.

HOURGLASS NEBULA

The hourglass shape of what, not surprisingly, has been called the Hourglass Nebula is produced by the expansion of a fast stellar wind within a slowly expanding cloud. The cloud is more dense near its equator than near its poles. The dying star, which is thought to be ejecting material and gas and giving light to the nebula, is actually off center. The intricate "etching" patterns may have been created by material shed from the star when it was a bit younger.

A DYING STAR
This image of a planetary nebula 8,000 light-years away shows a distinctive "hourglass" shape. It was created by the ejection of matter from a slowly dying star.

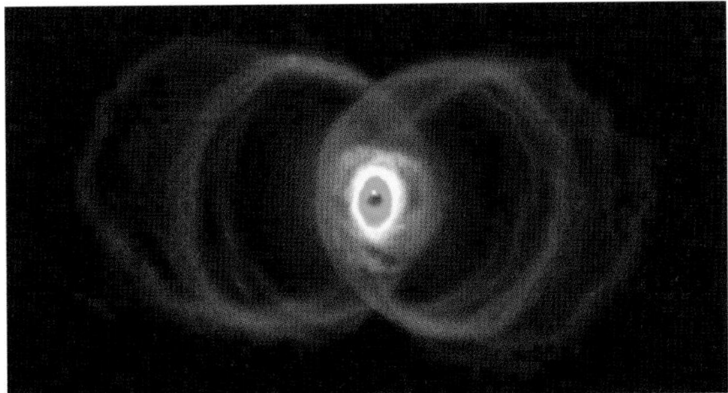

VIEWS FROM HUBBLE 2

The Hubble Space Telescope has returned spectacular images that cover a whole range of astronomical objects. These include the planetary nebula of a dying star, blue galaxies, and a possible "protoplanetary" system. The very common blue galaxies are not the same, familiar spiral shape when viewed by the HST. Instead they have a wide variety of shapes, which suggests that collisions between galaxies were more common in the early stages of the formation of the Universe.

Not all the images taken by the HST are spectacular, but they are nonetheless fascinating and sometimes quite historic. One image shows a protoplanetary disc, one of several seen in the Orion Nebula about 1,500 light-years away. It resembles a kind of interstellar flying disk. This disk of dust around a newborn star may be a planetary system in the making.

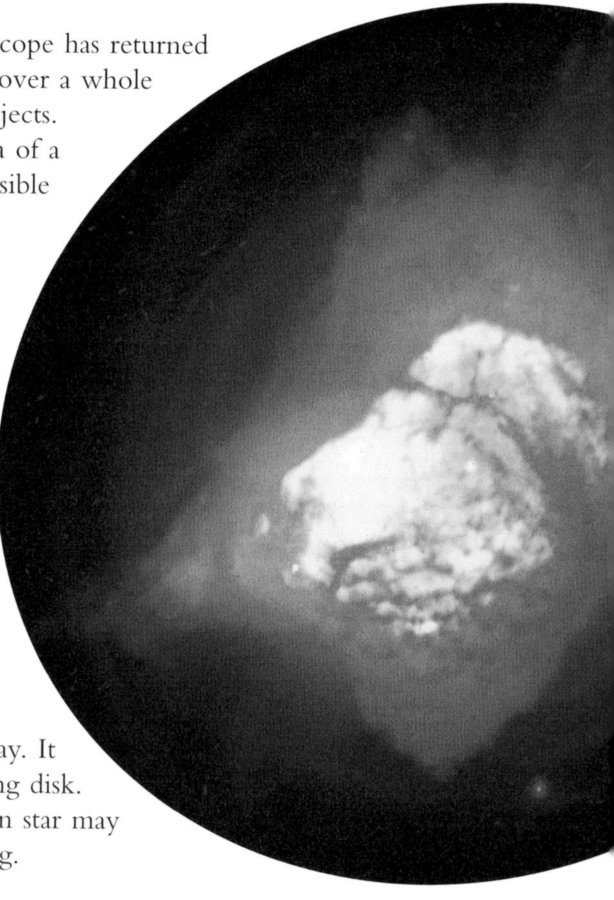

PROTOPLANETARY DISK
Does this image show a solar system in the making? It is a disk of dust and gas in the Orion Nebula. It is about 17 times larger than our own Solar System.

NGC 7027

The features of the NGC 7027 include faint, blue concentric shells surrounding the nebula, an extensive network of red dust clouds throughout the bright inner region, and the hot central white dwarf star that is visible as a white dot in the center. When a star like the Sun nears the end of its life, it expands to more than 50 times its original size, becoming a red giant. Then it gradually sheds its outer layers, exposing an extremely hot core. When this core cools off the star becomes a white dwarf.

DEATH OF A STAR
This picture shows the final death throes of a star about the size of our Sun. The planetary nebula NGC 7027 is about 3,000 light-years away, in the constellation Cygnus.

BLUE GALAXIES
According to astronomers, these faint blue galaxies are the most common objects in the whole Universe. Their distances from Earth range from 3 billion to 8 billion light-years away.

Infrared Observers

If we had infrared eyes instead of eyes that can see in the visible light wavelengths, the night sky would look totally different. We would see objects in the sky that are much cooler than our Sun. Infrared radiation passes through clouds of interstellar dust more easily than visible light does. This dust obscures much of the center of the Milky Way Galaxy from our eyes. Infrared observation satellites are providing a new view of the center of our Galaxy and of other, cooler objects too.

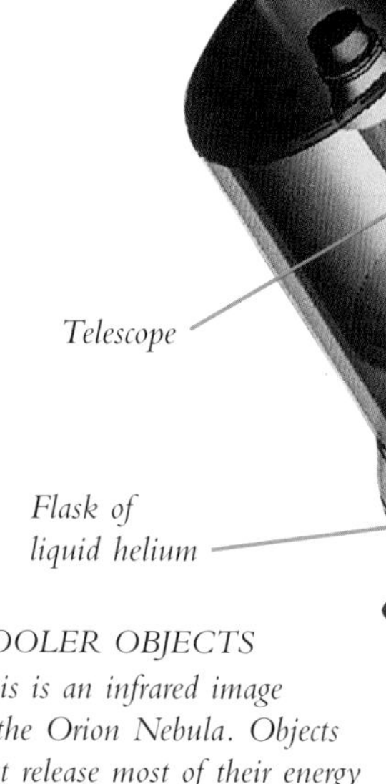

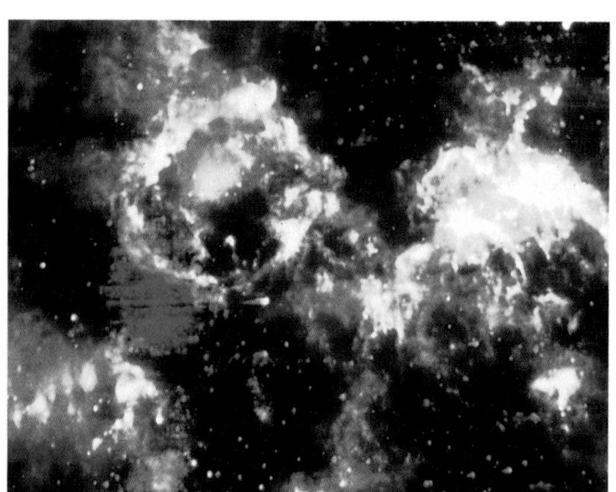

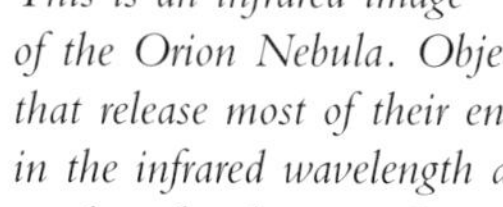

COOLER OBJECTS
This is an infrared image of the Orion Nebula. Objects that release most of their energy in the infrared wavelength are much cooler than our Sun, which has a temperature at its center of about 27 million degrees Fahrenheit (15 million degrees Celsius).

SENSITIVE DETECTORS
Infrared telescopes such as the one on board the Infrared Space Observatory (ISO) are fitted with detectors that are sensitive to infrared radiation.

THE ISO

The 2,498-kg Infrared Space Observatory was launched by an Ariane rocket in November 1995. It was placed into a 600-mile (1,000-km) by 43,000-mile (71,000-km) orbit around the Earth. This allowed the ISO to spend 16 hours a day observing the Universe from outside the Earth's radiation belts, which interfere with scientific observations. The craft was capable of observing specific objects for 12 hours continuously. The ISO was equipped with a cryostat, rather like a vacuum flask.
It contained 620 gallons (2,250 l) of liquid helium to cool the telescope.

Solar array and sunshield

SPACE SEARCH
The European Space Agency's Infrared Space Observatory (ISO) was launched in 1995. It is searching for super cold objects in the Universe.

Instrument module

These objects include potential planetary systems around stars, such as possible rings of what may be dust particles around the bright star Vega. Other objects include newly born stars, small, dim and dwarf stars, comets, asteroids, and quasars. It is vital to cool the telescope of an infrared spacecraft to keep it sensitive to infrared radiation. Liquid helium is commonly used to do this.

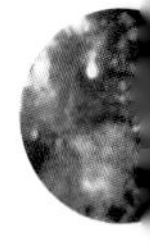

X-RAY OBSERVERS

X-rays cannot pass through the Earth's atmosphere, so astronomical observations have to take place beyond an altitude of 90 miles (150 km) above the Earth. The first observations were made by instruments that flew on up-and-down flights on sounding rockets. In 1962, one of these flights resulted in the first X-ray images of a star called Scorpius X-1, which is thought to be a double star.

Perhaps the most fascinating sources of X-rays are black holes. These may be the leftovers from exploded stars. They give off no visible light radiation, making them almost invisible, but their tremendous gravitational force results in very high X-ray emissions. Before matter drops into a black hole, it revolves like water going down a plug hole in a bath. This happens almost at the speed of light. Friction among the matter reaches temperatures of several million degrees, emitting X-rays that can be detected.

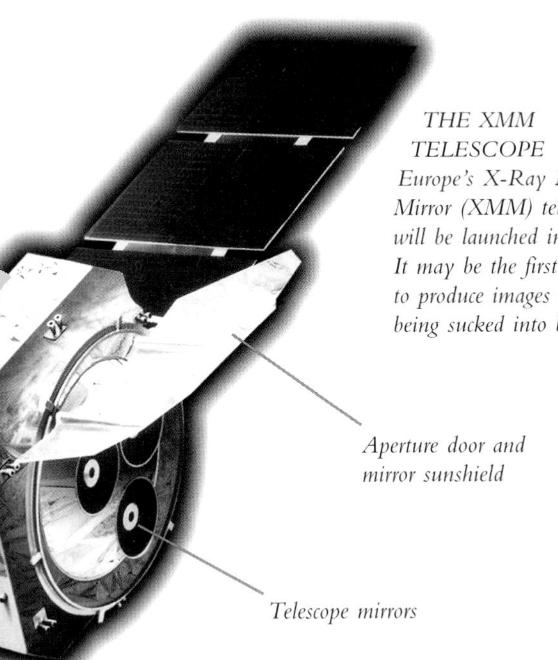

THE XMM TELESCOPE
Europe's X-Ray Multi-Mirror (XMM) telescope will be launched in 2000. It may be the first telescope to produce images of material being sucked into black holes.

X-RAY MULTI-MIRROR

The X-Ray Multi-Mirror (XMM) telescope is over 35 feet (11 m) long and weighs more than 4 tons. It will be placed into an orbit measuring 70,680 miles (114,000 km) by 4,265 miles (6,880 km), taking 48 hours to make one circle of the Earth. This orbit will enable the XMM to spend as much time as possible out of the belts of radiation circling the Earth. The telescope will be able to remain focused on one target for up to 40 hours. It will be accurate enough to detect perfectly the position of four melons perched on top of the White House, at a distance equal to that between Philadelphia and Washington.

X-RAY SOURCES
X-ray radiation is emitted strongly by very hot objects, such as neutron stars, and by those with very strong gravitational forces at work within them, such as black holes.

ULTRAVIOLET CRAFT

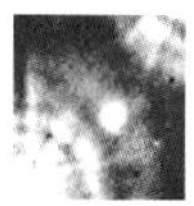

Ultraviolet (UV) radiation is absorbed by the Earth's atmosphere, which is a good thing since this radiation kills most living things. Astronomical observations of ultraviolet radiation are very important. Such observations have to take place high up above the atmosphere, starting at an altitude of about 20 miles (30 km). Important ultraviolet astronomy cannot take place until a craft is more than 150 miles (250 km) above the Earth, so a satellite is the main option for this type of astronomy.

The first UV satellites succeeded in making the first high-resolution map of the spectrum of a star in another galaxy, and the first ultraviolet view of a supernova. This type of astronomy focuses more on the area surrounding an object in space, rather than on the object itself. UV astronomy looks at the space between the stars, called the interstellar medium, which is invisible to the human eye. The interstellar medium consists of both hot and cold gases, and ultraviolet astronomy is a good way of mapping these regions.

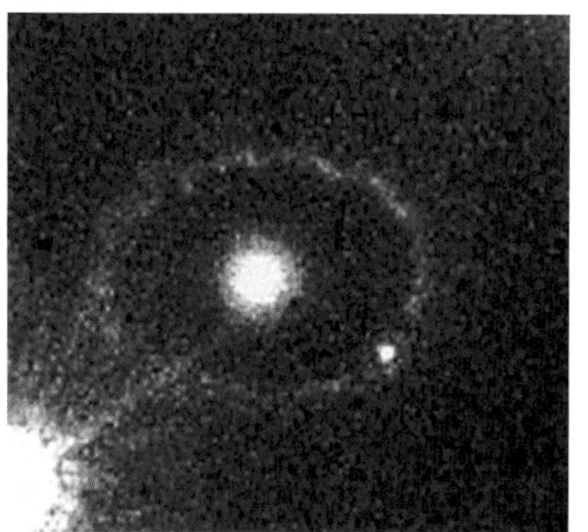

HYDROGEN IN SPACE
Only ultraviolet instruments on satellites can fully study molecules of hydrogen gas in space. Much of this gas is found in the interstellar medium in our Milky Way Galaxy.

UV EXPLORER
The Extreme Ultraviolet Explorer was launched in 1992. Its mission is to make the first all-sky survey in the extreme ultraviolet wavelength, which lies between ordinary ultraviolet rays and X-rays.

THE IUE

The International Ultraviolet Explorer (IUE) discovered that the planet Uranus shines brightly in ultraviolet. The reason for this was at first thought to be hydrogen gas trapped within the giant planet's magnetic field, called the magnetosphere. The US spacecraft *Voyager 2,* which passed the planet in January 1986, found no evidence of this. Astronomers now think that the ultraviolet glow is caused by particles from the Sun, which smash into molecules within the planet's upper atmosphere.

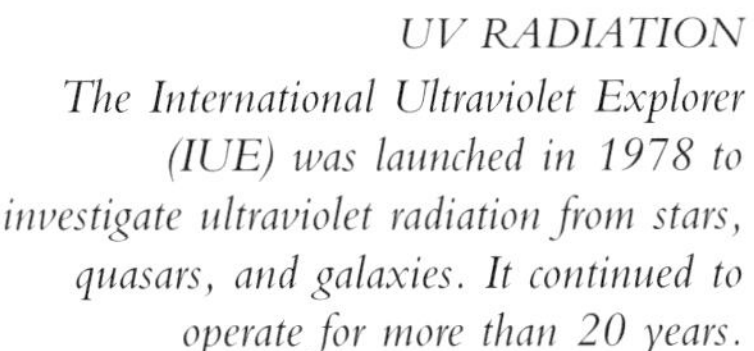

UV RADIATION
The International Ultraviolet Explorer (IUE) was launched in 1978 to investigate ultraviolet radiation from stars, quasars, and galaxies. It continued to operate for more than 20 years.

GAMMA AND COSMIC RAYS

Gamma rays are the most energetic form of electromagnetic radiation in the Universe. They are the by-products of some types of nuclear reactions and are produced by extremely violent reactions, such as those that occur in black holes, pulsars, and quasars. Like many other forms of radiation, gamma rays do not reach the Earth's surface, so gamma-ray astronomy did not begin until the space age.

The first gamma-ray observations were made by the American Orbiting Solar Observatory. In addition to studying the Sun, the observatory also found that gamma rays are very strong within about one third of the center of the Galaxy. Because gamma rays are so penetrating, gamma-ray astronomy even allows us to see through all the gas and dust in the Galaxy to discover what is happening at its center.

COMPTON GAMMA RAY OBSERVATORY
The Compton Gamma Ray Observatory was deployed from the space shuttle in 1990. It was the second of NASA's great observatories, after the Hubble Space Telescope.

REPAIR IN SPACE
The main antenna of the Compton Gamma Ray Observatory would not deploy before the observatory was ejected from the Space Shuttle. The Shuttle's astronauts had to go on a spacewalk to carry out the repairs.

COSMIC RAYS

Cosmic rays are the most powerful kind of energy in the electromagnetic spectrum. They are atomic particles that bombard the Earth from deep space. Some particles are created in supernova explosions, such as those that gave rise to the relatively close Crab Nebula and Vela Pulsar. When they hit the Earth's atmosphere, most cosmic rays break up into an "air shower" of secondary cosmic rays that reach the Earth's surface.
We are all exposed to cosmic rays. Strong cosmic rays can cause changes to living things on the Earth.

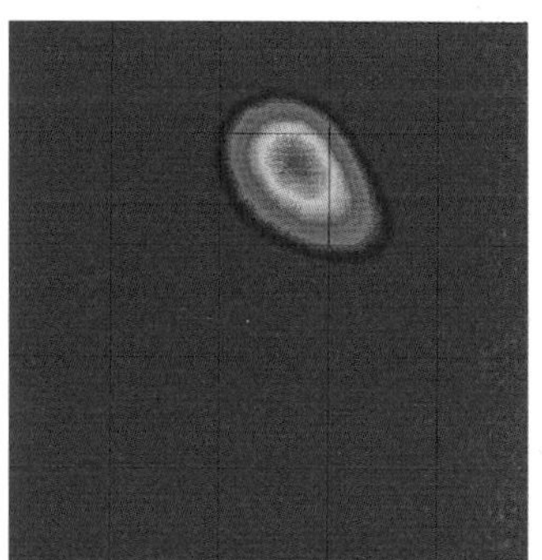

GAMMA-RAY IMAGE
Gamma rays are produced by the extremely violent reactions that take place at the centre of our Galaxy as well as in black holes, pulsars, and quasars.

LIFE OF A STAR 1

All stars appear to form out of clouds of interstellar gas and dust. The way in which a star dies, however, seems to depend on its size. When it became a star, our Sun was large and very bright. Gradually, as it reached middle age, it grew smaller but its temperature increased. The Sun is now at this stage and is called a main-sequence star. Eventually, it will begin to lose fuel and will increase

LIFE OF OUR SUN
The Sun was once about 50 times bigger than it is now. One day, as it dies, it will expand to about 400 times its present size, before turning into a white dwarf star, surrounded by a planetary nebula of gases.

Clumps of dust and gas pull together

A ball of gas forms in the center

After about 10 billion years, the star begins to shrink

Gravity squeezes the clumps

Heat from nuclear reactions makes the star shine

The star burns steadily

BIRTH OF A STAR
Most stars are formed out of an interstellar gas and dust cloud called a nebula. The gas compresses and creates incredibly high temperatures until a star is formed.

in size and brightness again, perhaps becoming a huge red giant about 100 times bigger than its present size. Eventually the Sun will lose material, which will form a planetary nebula.

The death of much larger stars is far more violent! A large star explodes and produces a super-bright supernova, shedding material in all directions. Sometimes the star blows itself to bits, but sometimes a tiny neutron star is left. In other cases the remaining core of the star is so massive that it continues to shrink, forming a black hole.

DEATH OF A STAR
Could this be the death throes of a star? This picture, taken by the Hubble Space Telescope, shows the star Eta Carinae. It became the second brightest star in the night sky in 1840, then faded to seventh magnitude.

ETA CARINAE

Eta Carinae became so bright at one time that it could have been classed as a supernova. It is no longer visible and exists in the clouds of dust and gas in the Homunculus Nebula, which is expanding at a rate of 300 miles (500 km) per second. Astronomers think that the star has not finished dying yet! Infrared images indicate that behind the clouds of dust and gas, Eta Carinae is still very bright and 100 times more massive than the Sun. It may explode in a real supernova and would appear brighter than Venus.

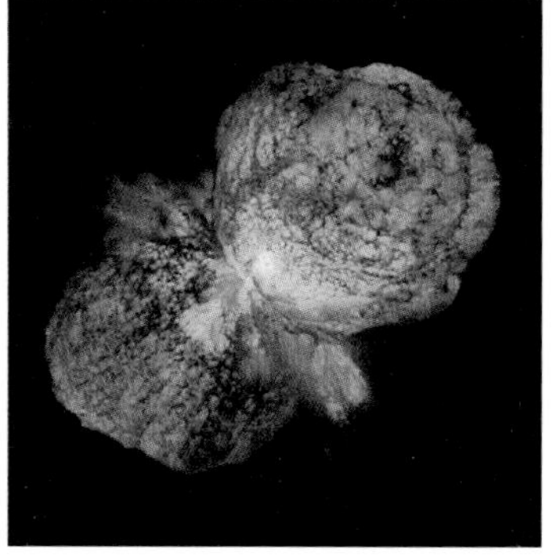

LIFE OF A STAR 2

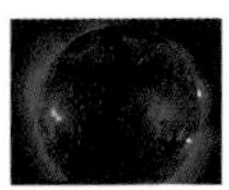

The development of stars is a very complicated process. There are many types of stars, each one at a different stage in its life. Generally, however, the development of a star depends mainly on its original size. Its color and brightness depend on its temperature. Blue stars are the hottest, while red stars are the coolest.

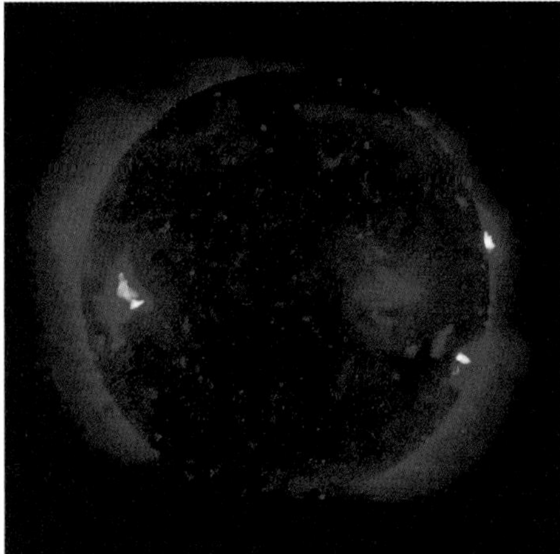

Usually, the color indicates the stage of the star's life. Most red stars are very old and have expanded to a huge size compared with their original size. The star Betelgeuse in the constellation Orion is one example.

Astronomers generally classify original stars in the light, middle, or heavyweight class. Red dwarf stars are common and are very faint, perhaps radiating one thousandth of the light that our yellow dwarf Sun does. The Sun is in the middle class but is still quite small compared with many stars in this category. The heavyweight stars usually develop into very hot blue giants. Over half the stars are in fact double stars.

A YELLOW DWARF STAR
Astronomers class the Sun as a yellow dwarf star in the middleweight class. It is likely to end its days as a red giant, before reducing in size to become a white dwarf.

True double stars are called binaries and should not be confused with two stars that happen to appear very close to each other in the sky but are in fact many light-years apart.

HEAVYWEIGHT STARS

Stars in the heavyweight class are expected to end their days as red supergiants, then transforming into either pulsars or black holes. There are a surprisingly large number of double stars in the Galaxy. They are also known as binary stars.

DOUBLE AND VARIABLE STARS

The double star Capella is the seventh brightest "star". It consists of two yellow stars, each about three times the size of the Sun. In some double stars, one of the stars passes in front of the other, blocking its light. This is a type of variable star, the best example being Algol in the constellation Perseus. Another type of variable star, for example Hercules X-1, is caused when an old star pulsates in and out, changing its brightness.

PULSARS

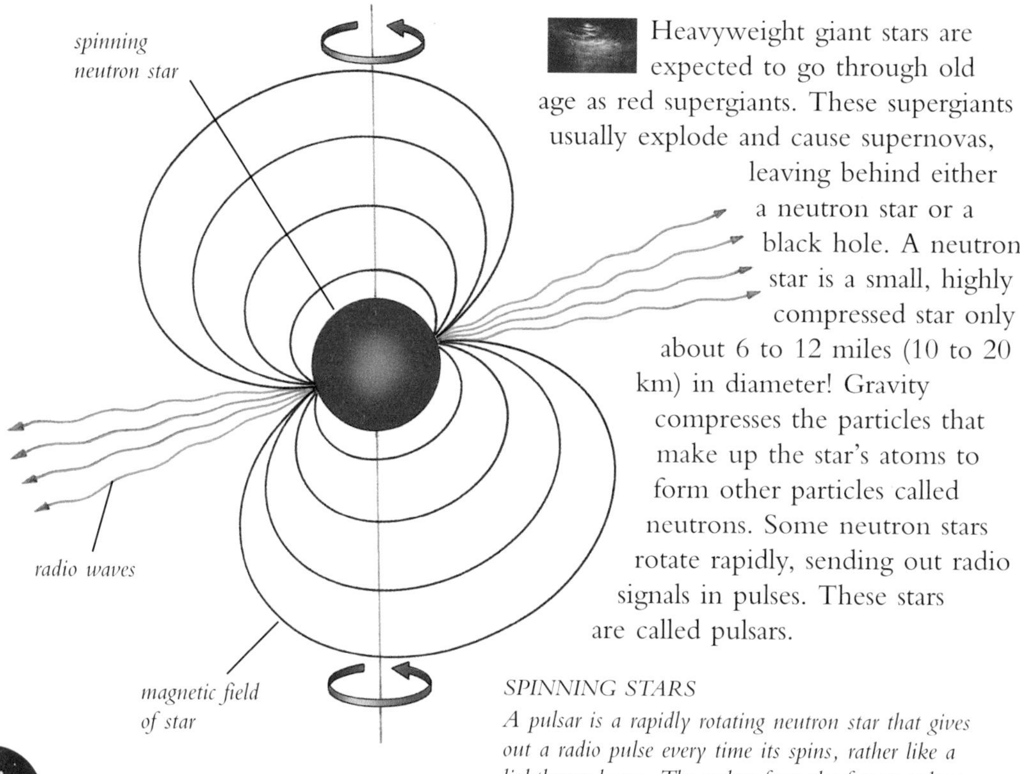

Heavyweight giant stars are expected to go through old age as red supergiants. These supergiants usually explode and cause supernovas, leaving behind either a neutron star or a black hole. A neutron star is a small, highly compressed star only about 6 to 12 miles (10 to 20 km) in diameter! Gravity compresses the particles that make up the star's atoms to form other particles called neutrons. Some neutron stars rotate rapidly, sending out radio signals in pulses. These stars are called pulsars.

SPINNING STARS

A pulsar is a rapidly rotating neutron star that gives out a radio pulse every time its spins, rather like a lighthouse beam. The pulses from the fastest pulsars are only a few milliseconds apart. Some pulsars also give off X-rays and gamma rays.

PULSAR IN CRAB NEBULA
The Crab Nebula's pulsar is shown in this image from the Hubble Space Telescope. The supernova that formed this nebula was seen from the Earth in A.D. 1054.

HISTORIC SUPERNOVA

Pulsars emit their signals in intervals varying from less than 1 second to about 3–5 seconds. The fastest pulsar — with a signal that flashes 30 times per second — is at the center of the Crab Nebula in the constellation Taurus. The Crab Nebula is the leftover material of one of the few supernovas that has been recorded in history. The supernova was seen by Arab and Oriental astronomers in July, A.D. 1054. It was so bright that it shone five times more brightly than Venus and could even be seen in the daytime.

Pulsars were first discovered in 1967 by radio astronomers in Cambridge, England. At first, the perfect timing of the signals created great excitement as many people believed they were artificial signals that might have been sent by other life forms. They were nicknamed the "little green men"! More than 300 pulsars have been discovered.

A DISTANT SUPERNOVA
This Hubble Space Telescope view shows a supernova that may one day leave behind a pulsing neutron star, or pulsar.

BLACK HOLES

Black holes are some of the most difficult phenomena in space to understand. When a star, usually a giant star, dies it may explode in a supernova. What is left over after the explosion depends on how much the remaining tiny star has been compressed. It may become a neutron star, or it may be compressed so much that space and time become distorted. The effect of this is to form a black hole whose gravitational pull sweeps up material like a vacuum cleaner.

A HOLE IN SPACE
A star ends its life by traveling out of our Universe, leaving behind only a hole in space. We can only imagine what a black hole might look like. Although black holes themselves are invisible to us, we can detect the X-rays given off by material falling into them.

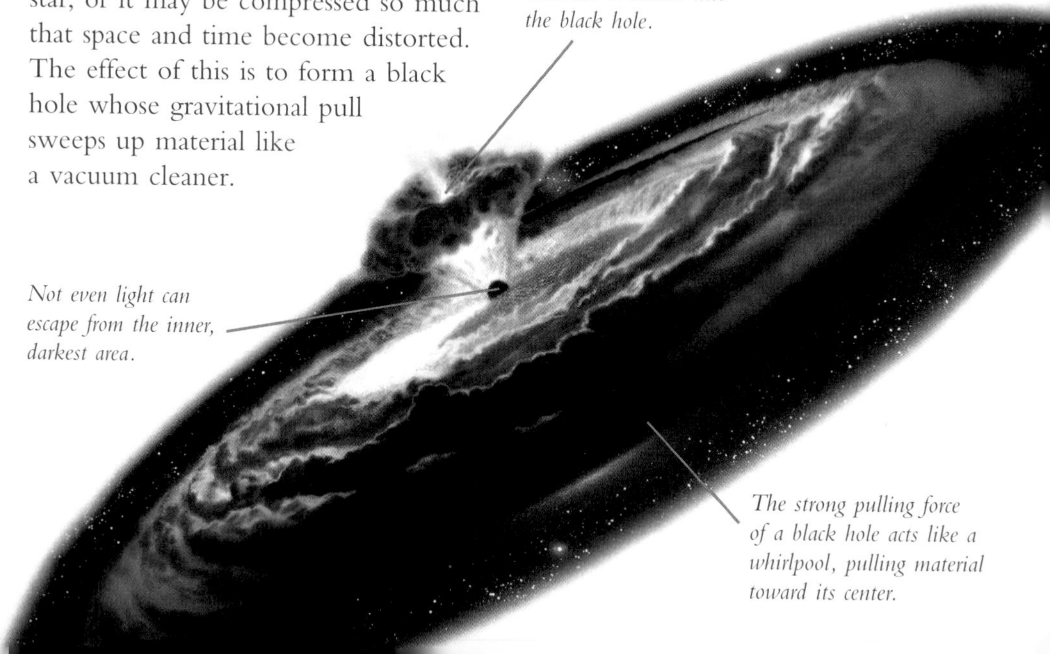

Material is sucked into the black hole.

Not even light can escape from the inner, darkest area.

The strong pulling force of a black hole acts like a whirlpool, pulling material toward its center.

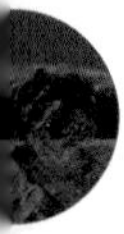

VIEW OF A BLACK HOLE
This is an artist's impression of what a black hole might look like to an observer on a planet circling a nearby star.

No amount of energy can escape from this pull — not even light, and so the collapsing star appears to be black. The whole event may seem to be almost instant in the "time" of the star, but it may have occurred over a period of millions of years. Although we cannot see black holes, we can detect their effect on nearby stars. As a star is slowly sucked into the black plughole it heats up, generating strong X-rays.

FINDING BLACK HOLES

X-ray satellites have detected a number of X-ray sources particularly near double stars. One source has been detected around the star Eta Cygni. The source is called Cygnus X-1, and it may contain a region with a black hole.
If a star about 10 times the size of the Sun were sucked into a black hole, the "opening" would only be about 20 miles (30 km) wide! Although black holes are thought to result from collapsing stars, they may also occur when galaxies collide.

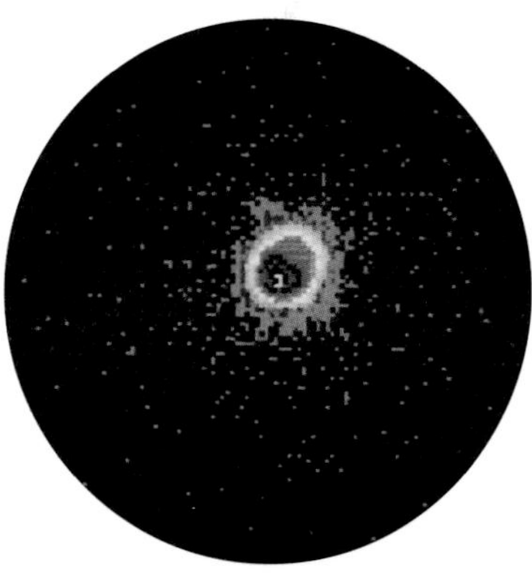

CYGNUS X-1
This view of Cygnus X-1 was taken in X-ray by the Rosat satellite. Astronomers think that it is the location of a black hole.

QUASARS

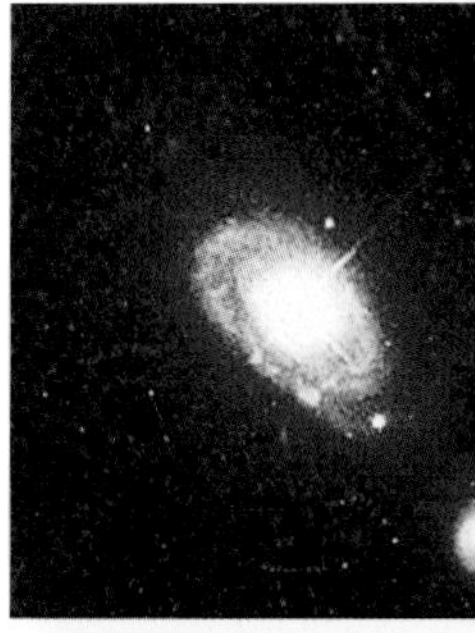

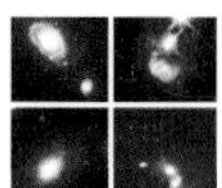

In 1963, astronomers discovered starlike points of light that apparently emitted more energy than an entire galaxy. These points of light are quasars, and they are a puzzle. It is a mystery how an object about one light-year across, and perhaps 100,000 times smaller than a galaxy, can emit so much energy. It is possible that the energy comes from very hot disks being heated up by the gravitational forces around a huge black hole. More recently, it has been suggested that quasars are very active central regions that cannot actually be seen inside galaxies. These galaxies are often called "active" galaxies. Quasars may also come from objects far more remote than observable galaxies, in which case they are the most powerful objects in the Universe.

HST VIEW OF A QUASAR
This quasar image from the Hubble Space Telescope (near left) is compared with a ground-based observation (far left). Since the first quasar was identified in 1963, astronomers have discovered more than1,000 other quasars.

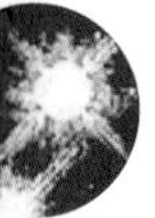

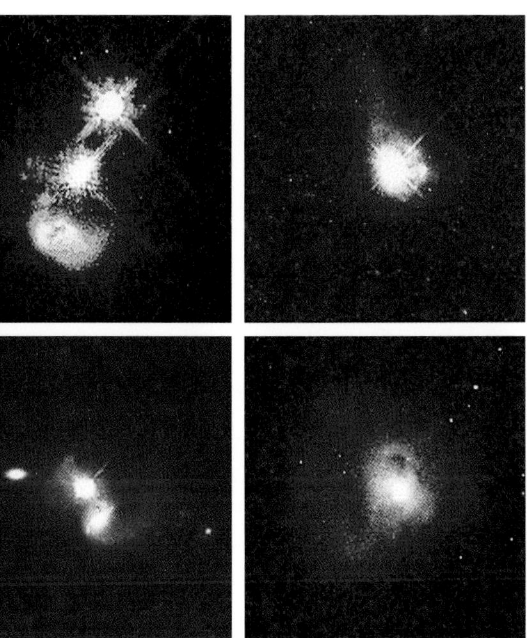

QUASAR IMAGES

The HST images (left) show: top left, a quasar 1.4 billion light-years away in a normal spiral galaxy; top center, an elliptical galaxy quasar about 1.5 billion light-years away; top right, a quasar that has apparently plunged vertically through the plane of a spiral galaxy 3 billion light-years away; bottom left, this quasar has merged with a bright galaxy 1.6 billion light-years away; bottom center, a tidal tail of dust and gas is seen beneath this quasar 2.2 billion light-years away; bottom right, glowing loops of gas around a quasar 2 billion light-years away provides evidence of merging galaxies.

QUASAR ENERGY

A quasar is a small and very bright object in distant space. It gives out the energy of hundreds of galaxies in a space not much bigger than our Solar System. These images taken by the Hubble Space Telescope show six different quasars seen in various galaxies billions of light-years away from the Earth. Astronomers think that quasars are probably the most distant objects thus far discovered in the Universe.

TYPES OF GALAXIES

A CLASSIC SPIRAL
A classic spiral galaxy has a very bright central region. The disk of very young stars is surrounded by a spiral of other stars, gas and dust.

Three main types of galaxies have been found in the Universe. Our Milky Way Galaxy is an example of a classic spiral galaxy. Its central disk is about 100,000 light-years in diameter and perhaps about 2,000 light-years thick. The galaxy has two arms spiraling around the central core. An elliptical galaxy is like a spiral galaxy without the arms. It looks a bit like a "splodge"of stars. Elliptical galaxies appear to be old galaxies. The largest are about 300,000 light-years in diameter. The third type of galaxy is an irregular galaxy. Examples of irregular galaxies are our Milky Way's small companion galaxies, the Large and Small Magellanic Clouds. These galaxies can only be seen in the night skies of the southern hemisphere.

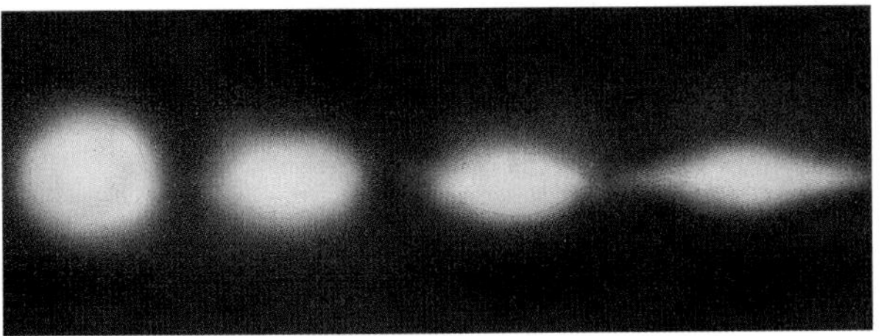

An elliptical galaxy is rather like a spiral galaxy without the arms. Elliptical galaxies come in different shapes, ranging from round globes and egglike shapes to squashed spheres. The light given off by an elliptical galaxy is strongest at the center, becoming fainter toward the outer edges.

WHIRLPOOL GALAXY

The first spiral galaxy was detected in 1845 by the third Earl of Rosse using his famous reflecting telescope at Birr Castle in Ireland. It is now known as the Whirlpool Galaxy, or M51. It is about the same size as our Milky Way Galaxy. The Whirlpool Galaxy is in the constellation Canes Venatici. It is not a very prominent constellation, consisting basically of two stars below the "bent handle" of Ursa Major.

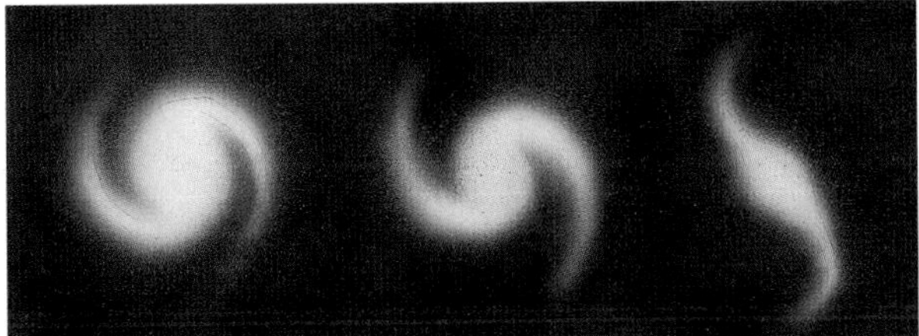

A spiral galaxy is a flattish disk with a large bulge at the center. Spiral arms coil outward directly from the bulge. The galaxy resembles a spinning Catherine wheel firework. Within the galaxy there are dark clouds of gas and dust. All spiral galaxies spin around.

A barred spiral galaxy is similar to a spiral galaxy but it has a bridge of stars across the centre. This bridge joins together the inside ends of the two spiral arms. Most of the stars are in the galaxy's spiral arms.

BIG BANG

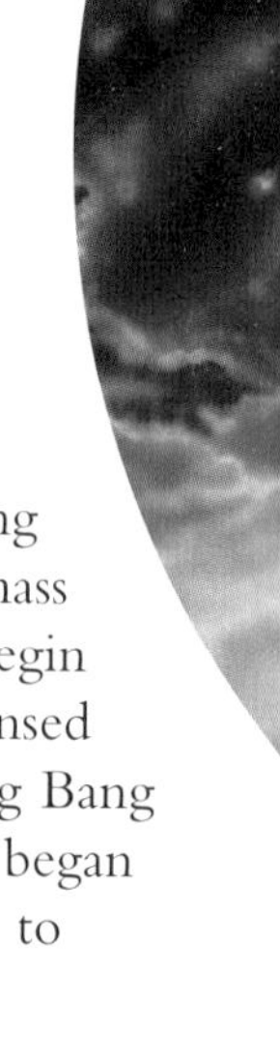

Scientists have three main theories about the creation of the Universe: the steady state theory, the "Big Bang" theory and the pulsating Universe theory. Whichever process was involved, this will not explain once and for all who made the Universe or whether it just "happened". The steady state theory says that the Universe has always been expanding, that new material is constantly being created and there is the same amount of matter in the same place.

According to the pulsating Universe theory, all matter is flying apart from a heavily compacted mass and will eventually slow down, begin to contract and become so condensed that it will explode again. The Big Bang theory suggests that the Universe began in an explosion and will continue to expand forever and ever.

THE BIG BANG THEORY
Some scientists believe that the Universe began in a "Big Bang", and that it will continue to expand indefinitely while no new matter is being created. According to this theory, the early Universe was very hot and dense, with all its matter and space packed into a very small area.

CREATING A UNIVERSE

Not all scientists agree with the theory that the Universe is the result of an enormous explosion. They do not believe that it just "happened." Many believe that a god or gods created the heavens and the Earth. If the Universe just "happened," what made its extraordinary variety and beauty? Does one of today's digital televisions just appear in the living room? No! It has been carefully designed, manufactured, and delivered!

THE EXPANDING UNIVERSE
Most astronomers believe that the Universe was created in a "Big Bang" explosion about 17 billion years ago. At first the Universe was quite small, but as it cooled down it expanded. Astronomers believe that it is still expanding today.

Is There Life Out There?

Is there life in the Universe? The answer is yes — there is life on the Earth. The Earth is unique. It is the only planet — and the only place in the Universe — where we know that life definitely exists. This life is not just small microbes. It is a life of incredible variety. Millions of different mammals, birds, reptiles, insects, fish, and plants.

There are about 275,000 different types of flowering plants and trees on the Earth, all of them distinctly different. More than 1 million insects have already been discovered.

The Earth is just the right distance from the Sun to support life. It has lots of fresh water in liquid form. The Earth also has a perfect atmosphere to support life. It is mainly made up of nitrogen and oxygen. The Earth is tilted in such as way to cause seasons. These seasons cause a range of weather conditions that help to support all kinds of living things.

ALIEN BEINGS

As there is life on the Earth, some scientists believe that it could exist elsewhere in the Universe. They think that there may even be alien civilizations. People have had many imaginative ideas about what aliens from elsewhere in the Universe might look like — but no one can know for sure!

LIFE ON EARTH
Some scientists think that life on Earth may have begun when it was "seeded" by microbes. These microbes might have come from passing comets or other objects in the early history of the Universe.

OUR LIFE

Most scientists believe that the extraordinary range of life on Earth just happened to start 3.5 billion years ago. It began with simple microscopic living cells in a bubbling pond of water. Many scientists say that water is the most important requirement for life to form. Because traces of water have been found in other parts of the Solar System, people have assumed automatically that life must therefore exist elsewhere. Many other people believe that a god or gods created the Earth and its huge variety of living things.

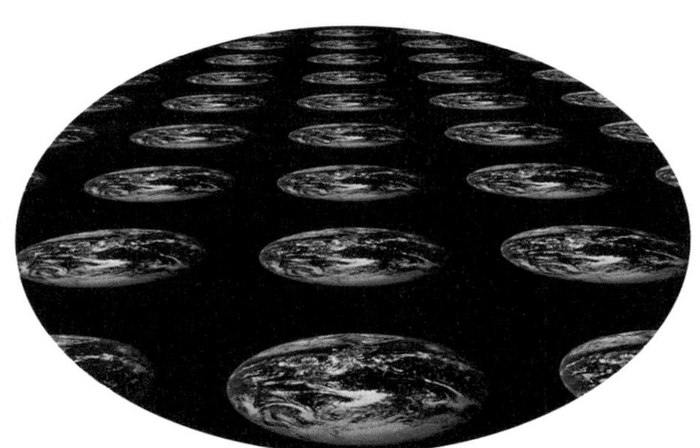

LIFE IN THE UNIVERSE
We know that life exists on a tiny planet (the Earth) orbiting a little star (the Sun) in a part of the Milky Way Galaxy. Some scientists think that there may be other "earths" with life forms elsewhere in the Universe.

THE SOLAR SYSTEM

The Sun, together with the family of nine planets, comets, asteroids, and other bodies that travel round it, make up our Solar System. It was probably formed from a cloud of gas around 4.6 billion years ago. At the very heart of the Solar System lies the Sun, providing the heat and light that make life possible on the planet Earth.

Spacecrafts have visited all of the planets, with the exception of Pluto, the smallest and most distant planet. They have revealed a wealth of

different features on the planets, including mountains and valleys, craters and active volcanoes, colored ring systems, poisonous atmospheres, and polar ice caps. Earth is still the only planet in the Solar System where we know that life definitely exists, although scientists are continuously looking for signs of life on all of the other planets.

THE SUN AND FAMILY

Scientists believe that the Sun was born out of a cloud of hot gas and dust. As it started to become a fully fledged star, the remaining material was left orbiting the Sun at high speed. Specks of dust began to join together and form tiny rocks, which fused together gradually forming large bodies surrounded by clouds of gas. These bodies are the planets. The Solar System consists of nine known major planets: Mercury, Venus, Earth, Mars, Jupiter, Saturn, Uranus, Neptune, and Pluto.

Other material was left orbiting the Sun when the Solar System was formed. It includes the asteroids, a network of smaller minor planets most of which are orbiting the Sun between Mars and Jupiter. Other left-over material includes comets and small and large pieces of rock, called meteoroids, which travel in random orbits around the sun.

THE EARTH IS FORMED

The Sun was formed out of a spinning cloud of gas and dust. The rest of the Solar System is thought to be made up of the remaining material left over after the formation of the Sun. The gas and dust attracted other material which slowly developed into the planets, including the Earth.

SATURN, THE RINGED PLANET
The planet Saturn is considered to be the "star" of the Solar System. It has a well-formed system of rings and a number of moons. The other outer planets have much smaller ring systems.

THE GAS PLANETS

Saturn is one of the four large gaseous planets in the Solar System. The others are Jupiter, Uranus, and Neptune. These large planets, which have small solid cores, are made mainly of frozen gas. Saturn's rings may consist of the original material left over when the Solar System was formed. Another, more likely theory is that the rings are the remains of a moon that came too close and then disintegrated under the forces of the planet's gravity. Jupiter, one of the other gas planets, is so large and has such a huge atmosphere that many astronomers believe it almost became a star.

HALLEY'S COMET
Comets are made of material left over when the planets were formed. They consist of dust and frozen gases. Comets shine in the sky because they are lit up by the Sun as they pass close to it.

THE SUN

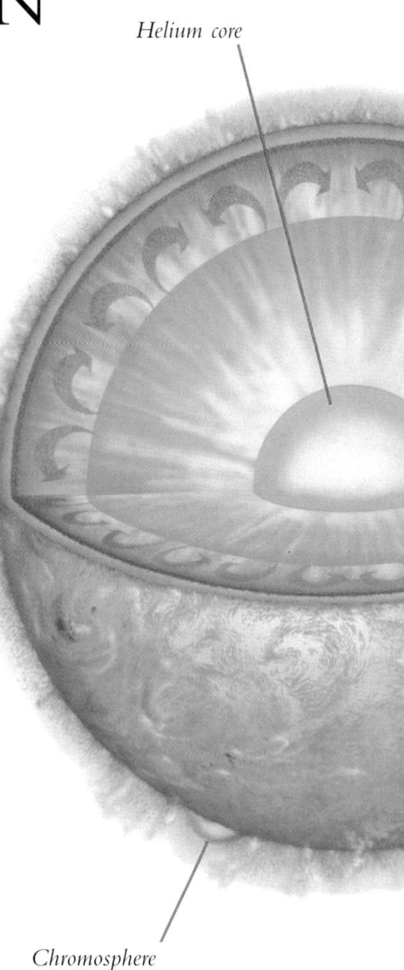

The Sun is 93 million miles (150 million km) away from the Earth. It is 109 times the size of the Earth and contains 99.9 percent of the mass of the whole Solar System. The Sun's center is like a huge nuclear furnace in which the temperature and pressure inside are so high that they set off atomic reactions. Here, atoms of hydrogen fuse together to form helium, and the energy produced at the core radiates out toward the surface.

The surface of the Sun is called the photosphere.The temperature here ranges from 7,740 degrees F (4,300 degrees C) to 16,200 degrees F (9,000 degrees C). The photosphere provides most of the light that comes from the Sun. The upper level of the photosphere, the chromosphere, is a stormy region of very hot gases. Here, the temperature has risen to nearly 2 million degrees F (1.1 million degrees C). The chromosphere is about 10,000 miles (16,000 km) thick. Above it, the Sun has a halo of even hotter gases called the corona.

OUR NEAREST STAR

The Sun is the nearest star to the Earth. It is a main-sequence yellow dwarf star and is rather insignificant compared with the much larger stars in the Milky Way. Light from the Sun takes 8 minutes, 17 seconds to reach the Earth, compared with the 4.3 light-years it takes to come from the next nearest star, Proxima Centauri. The center of the Sun is a huge nuclear furnace with a temperature of 27 million degrees F (15 million degrees C).

Hydrogen layer

Solar flare

Photosphere

Sunspot

THE SOLAR WIND

The outer layers of the corona are made up of hot gases blowing off from the Sun. This stream of gases is called the solar wind. It flows away from the Sun and through the Solar System. Sometimes, when the solar wind meets up with the Earth's upper atmosphere, or ionosphere, it causes magnetic storms and radio interference. The solar wind also causes auroras, the red and green glowing lights that are visible in the night sky in the extreme parts of the northern and southern hemispheres.

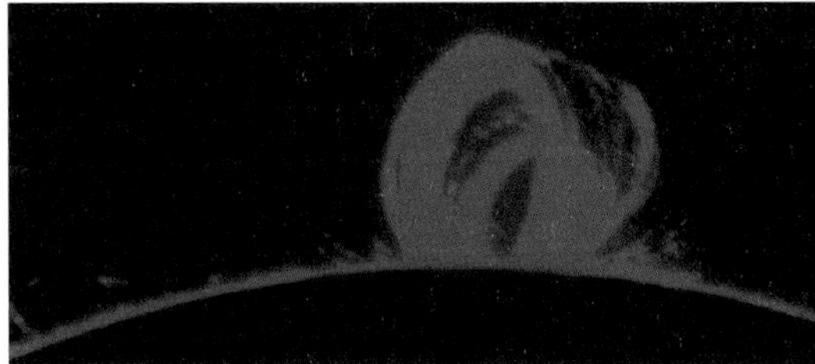

SOLAR ACTIVITY

Bright arches of hot glowing gas erupt from the surface of the Sun. These arches are called prominences. They may reach as far as 20 miles (30 km) above the Sun's surface. Some have a loop shape (above), while others are like a curtain of gas.

MERCURY

Mercury, the nearest planet to the Sun, is a small rocky body with a diameter of 3,025 miles (4,880 km). The planet takes just 88 Earth days to travel once around the Sun at a distance of between 43 million miles (69 million km) and just 30 million miles (49 million km).
The average distance of Mercury from the Sun is 35 million miles (57 million km), which is about half as close to the Sun as the Earth is.

The temperature in the intense sunlight on Mercury is hot enough to melt lead at midday. Mercury rotates very slowly, once every 58 Earth days. Because of this, each day on Mercury lasts 176 Earth days. The nighttime temperature is –320 degrees F(–180 degrees C). There is no atmosphere on Mercury and it would be impossible to live there.

MERCURY FACT FILE

Diameter: 3,025 miles (4,880 km)

Average distance from Sun: 35.9 million miles (57.9 million km)

Length of a year: 88 Earth days

Number of moons: 0

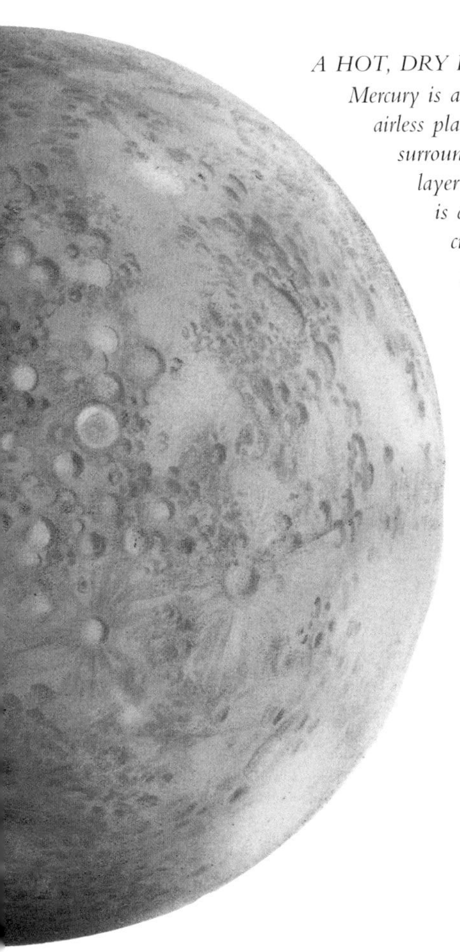

A HOT, DRY PLANET
Mercury is a very hot, dry, and airless place. The planet is surrounded by a very thin layer of gases. Its surface is covered with many craters, which were probably formed when meteorites or comets crashed into the planet.

WATCHING MERCURY

Because of its closeness to the Sun, Mercury never strays far from the Sun in the sky and so it is difficult to see. Occasionally, however, the small planet passes in front of the Sun as seen from Earth. This is called a transit. The next transit will occur on November 14, 1999. Observers who have the correct equipment to look safely at the Sun will see Mercury as a small, slow-moving dot.

MERCURY IN VIEW
Nobody knew what Mercury looked like until a space probe flew past the planet and took close-up pictures. Mercury looks very much like our own Moon. Its surface is covered with craters and mountains.

VENUS

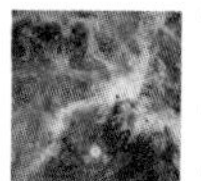

Venus is surrounded by thick clouds of carbon dioxide gas, which trap the Sun's heat, help to create a greenhouse effect on the planet. The trapped heat raises the temperature on the surface to 855 degrees F (475 degrees C), despite only about 2 percent of the Sun's light reaching the surface. The atmospheric pressure on Venus is 90 times greater than on Earth. Also, it rains sulfuric acid there. So if you stood on Venus, you would be boiled, squashed, and dissolved in one go!

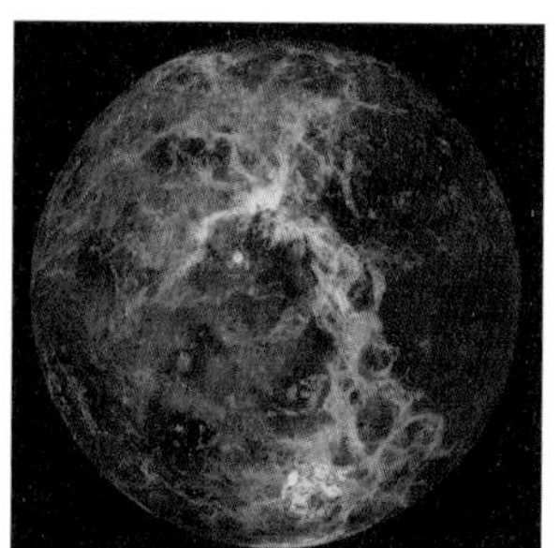

EARTH'S 'TWIN' PLANET
Venus is almost the same size as the Earth. It used to be known as Earth's sister or twin until people realized what was hidden beneath its thick sulphuric clouds. The surface of the planet is very hot and dry.

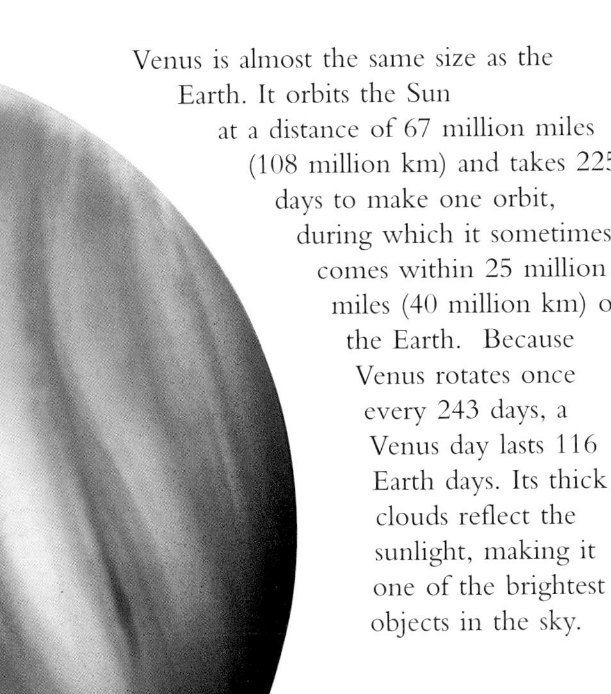

Venus is almost the same size as the Earth. It orbits the Sun at a distance of 67 million miles (108 million km) and takes 225 days to make one orbit, during which it sometimes comes within 25 million miles (40 million km) of the Earth. Because Venus rotates once every 243 days, a Venus day lasts 116 Earth days. Its thick clouds reflect the sunlight, making it one of the brightest objects in the sky.

BRIGHT PLANET
The thick clouds that make up Venus's atmosphere reflect the sunlight brightly. It is therefore the brightest object in the Earth's skies apart from the Sun and the Moon. The planet's surface is covered with mountains, craters, and volcanoes, some of them bigger than Mount Everest, the highest mountain on the Earth.

VIEWS OF VENUS

Before the surface of Venus could be seen in telescopes, and before spacecraft went close to it, Venus was seen as a rather romantic world. It was viewed as lying beneath a veil of clouds that reached a height of about 60 miles (100 km) above the surface. People assumed that Venus, like Earth, had water. Because the planet is closer to the Sun than we are, many people, even in the 1960s, thought that the planet's surface was like a tropical jungle.

VENUS FACT FILE

Diameter: 7,500 miles (12,100 km)

Average distance from Sun: 67.1 million miles (108.2 million km)

Length of a year: 225 Earth days

Number of moons: 0

THE EARTH

The Earth orbits the Sun once every 365 days, or year. As it orbits the Sun, the Earth is rotating at a speed of 1,030 miles (1,660 km) per hour. It makes one rotation every 24 hours, or day. The Earth travels through space at a speed of almost 20 miles (30 km) per second, and in a year it travels a total distance of 595 million miles (960 million km). The Earth has one moon which orbits at an average distance of 238,330 miles (384,400 km).

THE EARTH FROM SPACE
When seen from space, the Earth is the most beautiful and brightest planet. About three quarters of its surface are covered by very reflective oceans of water. If we could see the Earth from another planet, it would look like a bright bluish star.

SPACESHIP EARTH

The Earth is like a spaceship. It travels through space at a speed of almost 20 miles (30 km) per second. Although the Earth has a diameter of 7,908 miles (12,756 km), it has a very thin crust which is only 20 miles (32 km) thick.

The Earth has an atmosphere that is rich in oxygen (21 percent) and nitrogen (78 percent). This atmosphere protects the Earth from deadly radiation from the Sun. About 70 percent of the Earth's surface is covered in liquid water in the form of the seas and oceans.

EARTH STATISTICS

• The temperatures on Earth vary from about – 128 degrees F (–89 degrees C) to about 136 degrees F (58 degrees C).

• The deepest point on the Earth is the Marianas Trench, which lies 35, 830 feet (10,924 m) below the Pacific Ocean.

• The highest point on the Earth is Mount Everest (29,021 feet/8,848 m).

•The North Pole does not point directly up because the Earth's axis is slightly tilted (by about 23.5 degrees).

EARTH FACT FILE

Diameter: 7,908 miles (12,756 km)

Average distance from Sun: 93 million miles (150 million km)

Length of a year: 365.25 Earth days

Number of moons: 1

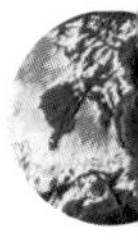

Our Moon

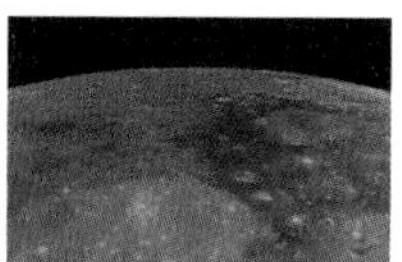

The Moon is about one third the size of the Earth. The Moon–Earth system is sometimes referred to as a "double planet." The Moon orbits the Earth every 27.5 days. Its surface is very dark because only about 7 percent of the Sun's light is reflected by it. The temperature on the Moon varies from 220 degrees F (105 degrees C) in the bright sunlight to minus –247 degrees F (–155 degrees C) in the shade. The pull of gravity on the Moon is only one-sixth of that on the Earth. When you look at the Moon you can see the famous "Man in the Moon" face. This "face" is created by the lighter areas on the Moon's surface, which are covered by craters and mountains, and by the darker areas which are flat, wide plains. These plains were mistaken for "seas" by early astronomers, which explains why they have such names as Sea of Tranquillity.

PHASES OF THE MOON
When the Moon is between the Sun and the Earth its far side is lit up, so we cannot see the side that faces us. When the Moon has moved farther round in its orbit we can see a crescent shape (far left), which gets bigger until a full Moon (center) appears.

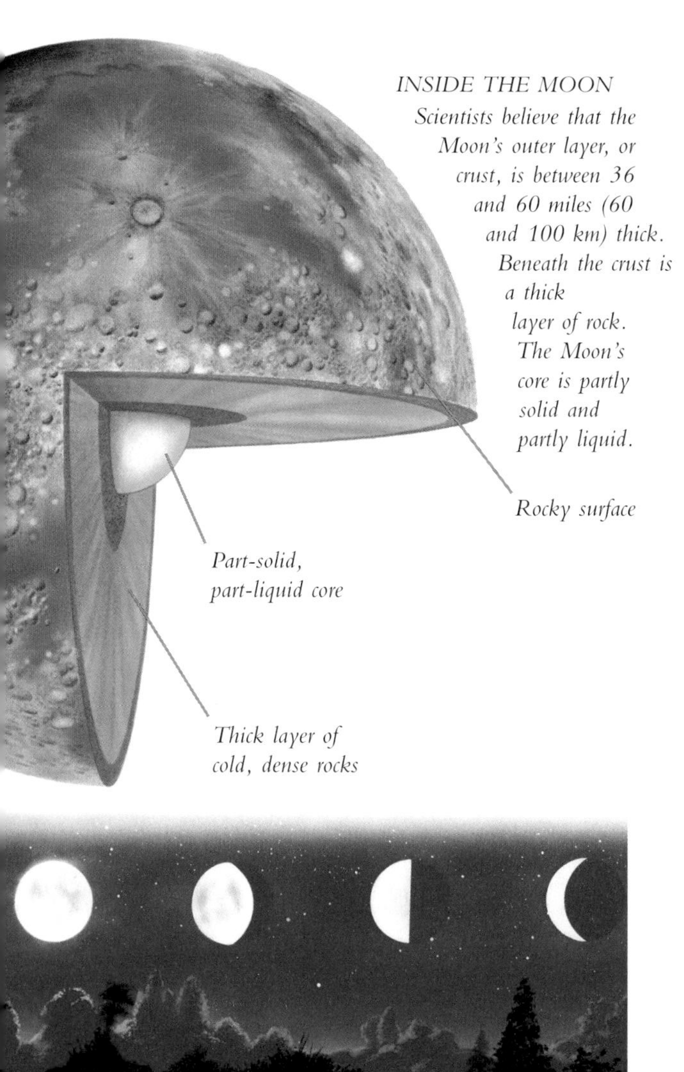

INSIDE THE MOON
Scientists believe that the Moon's outer layer, or crust, is between 36 and 60 miles (60 and 100 km) thick. Beneath the crust is a thick layer of rock. The Moon's core is partly solid and partly liquid.

MOON CRATERS

Many large craters on the Moon were named for famous people, especially famous astronomers. The most spectacular craters are those with "rays", which are made of material that was ejected from the craters when they were formed by the impact of meteorites. The most famous crater is called Tycho and can be seen clearly with the naked eye. The brightest crater is called Aristarchus. The largest crater on our side of the Moon is called Clavius. It has a diameter of 144 miles (232 km).

MOON FACT FILE

Diameter: 2,155 miles 3,476 km

Average distance from Earth: 238,328 miles (384,400 km)

Time taken to make a complete orbit of Earth: 27.5 Earth days

Average speed of orbit: 2,300 mph (3,700 km/h)

Mars

Mars has always excited astronomers because it is the only planet that can be observed clearly from Earth using quite small telescopes. Mars orbits the Sun every 687 days — the length of a Martian year. The length of a day on Mars is similar to that on Earth — 24 hours and 37 minutes.

Unlike the Earth, the atmosphere on Mars is 95 percent carbon dioxide. A maximum Martian temperature of –20 degrees F (–29 degrees C) makes Mars as cold as the coldest place on Earth. The atmospheric pressure there is just 1 percent that of the Earth's, so Mars would not be able to support life as we know it. Mars has two moons, called Deimos and Phobos. They are huge chunks of rock shaped like pockmarked potatoes. Phobos, the largest and closest one, is 17 miles (27 km) long and 12 miles (19 km) wide.

VIKING VISIT TO MARS

This image of the surface of Mars was taken by a Viking *spacecraft that landed on the planet in 1976. The Martian surface is covered with reddish sand, dotted with various different sizes of rocks. The sand looks as though it may have been deposited by running water, which has also smoothed the rocks.*

THE RED PLANET

Mars is sometimes called the Red Planet because it has a reddish surface, which indicates the presence of iron oxide in the soil. On the Earth, iron oxide is known as rust. Mars has a very active environment with dust storms, fog, frost, and polar ice caps made of dry carbon dioxide and water ice. Mars has a spectacular landscape of volcanoes, craters, and canyons.

LIFE ON MARS

Early observations by telescope showed the dark areas on Mars changing shape during the year. The planet's polar ice caps were also sighted. People then started to think that water from the melting ice caps was helping to cultivate large areas of vegetation during the summer. One astronomer thought he could see lines on Mars, and the idea took off that there were Martians who had built irrigation canals! Mars then became the subject of many "sci-fi" stories, such as the famous H.G.Wells story *War of the Worlds*.

MARS FACT FILE

Diameter: 4,208 miles (6,787 km)

Average distance from Sun: 141 million miles (227.9 million km)

Length of a year: 687 Earth days

Number of moons: 2

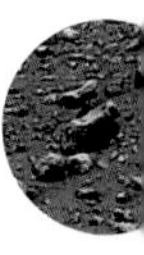

Asteroids

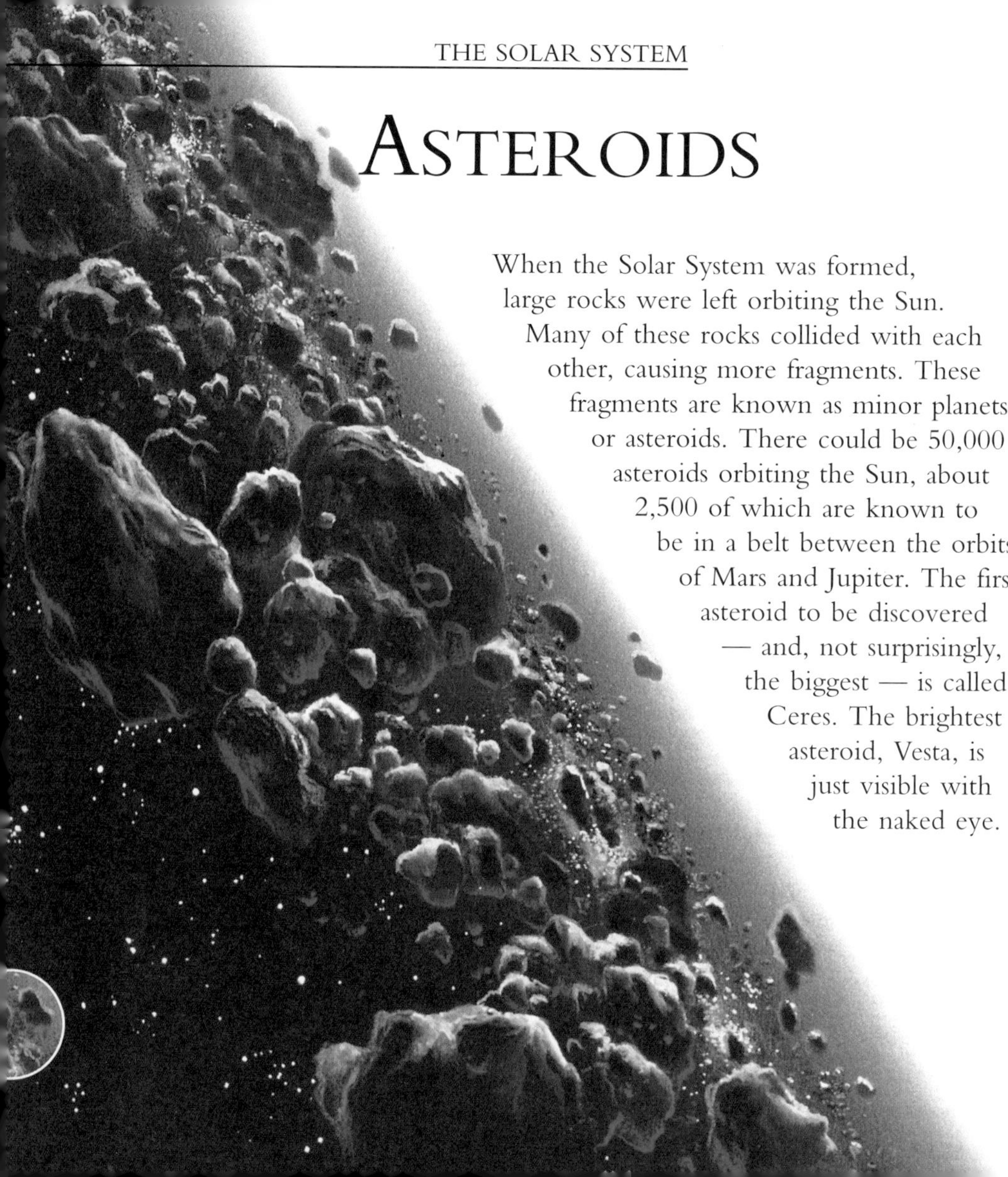

When the Solar System was formed, large rocks were left orbiting the Sun. Many of these rocks collided with each other, causing more fragments. These fragments are known as minor planets, or asteroids. There could be 50,000 asteroids orbiting the Sun, about 2,500 of which are known to be in a belt between the orbits of Mars and Jupiter. The first asteroid to be discovered — and, not surprisingly, the biggest — is called Ceres. The brightest asteroid, Vesta, is just visible with the naked eye.

ASTEROID GASPRA
Most asteroids, like this one called Gaspra, are in orbit in an area between Mars and Jupiter — the "asteroid belt." The largest asteroid, called Ceres, is 580 miles (936 km) wide.

There are other asteroids in elliptical orbits around the Sun. These come close to the Earth in their orbits, then travel far out into the deepest parts of the Solar System. They are called "Sun-grazers". It is quite possible that asteroids have hit the Earth, causing extensive damage to its surface and the atmosphere. An asteroid impact might have been the cause of an explosion in 1908 in Siberia. It had a force equivalent to that of a 13-megaton nuclear bomb.

ROCKY BODIES
Thousands of asteroids orbit the Sun. These rocky bodies are the bits and pieces of material that seem to have been left over when the Solar System was formed. Although the largest asteroid is about 600 miles (1,000 km) wide, most of the known asteroids are less than 12 miles (20 km) in diameter.

"SUN-GRAZERS"

The "Sun-grazer" Eros is an elongated asteroid. One of the most famous Sun-grazing asteroids is Icarus. It is just 0.9 miles (1.4 km) wide and comes to within 18 million miles (29 million km) of the Sun during its 1.1-year orbit. It actually glows red from the Sun's heat. In 1968, Icarus could be seen like a faint star traveling across the sky as it passed just 4 million miles (6.4 million km) from the Earth.

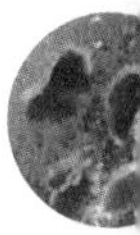

JUPITER

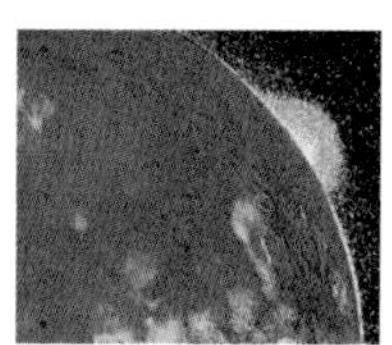

When the Solar System and the Sun were formed, Jupiter almost became another star. It is the largest planet, with a diameter of 88,540 miles (142,800 km). Jupiter is a huge ball of gases, including hydrogen, helium, ammonium, hydrogen sulfide and phosphorus. Bands of clouds of these gases swirl around Jupiter, which rotates quicker than any other planet — taking just 10 hours.

The colorful surface of Jupiter is dominated by the Great Red Spot, a swirling hurricane of gases where winds reach speeds of 21,700 miles per hour (35,000 km/h). The Great Red Spot could swallow up the Earth several times. Jupiter has at least 16 moons, and there could be many more that are too small to see. Its four major moons, from largest to smallest, are Ganymede, Callisto, Io, and Europa. Jupiter also has a small ring system that cannot be seen from Earth. It was discovered by a spacecraft sent to explore the giant planet.

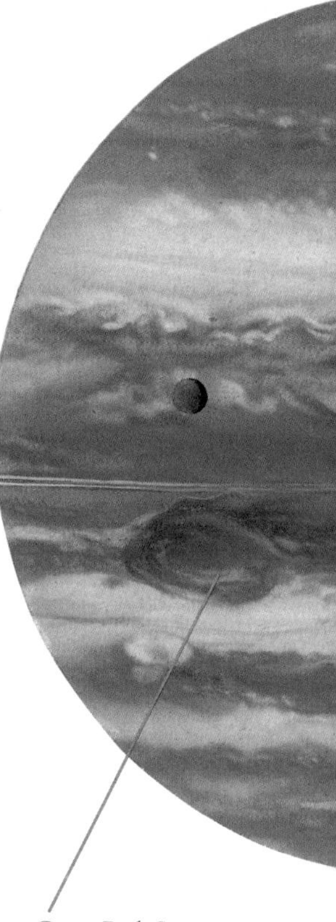

Great Red Spot

THE LARGEST PLANET

Jupiter is the largest planet in the Solar System. Astronomers believe that it is a failed star which, had it been a bit bigger, could have helped to form a double-star system. Galileo made the first recorded observations of Jupiter through a telescope and saw four small specks orbiting the planet. These moved position each day. They are known as the Galilean moons.

Faint rings made of dust particles

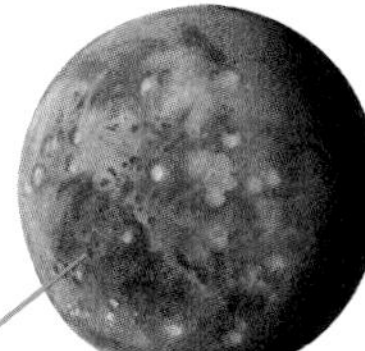

Galilean moon

JUPITER'S MOONS

The surface of Io is a dramatic world of orange, yellow and white deposits of sulfur and sulfur dioxide, with active volcanoes. Europa is like a huge ice pack, streaked with cracks. Underneath the ice pack there may be oceans of water, which some scientists think could be a source of basic living cells. Ganymede, the largest of the Galilean moons, has a surface that looks like a badly cracked eggshell, and Callisto has a pockmarked surface that resembles the skin of an avocado pear.

JUPITER FACT FILE

Diameter: 88,540 miles 142,800 km)

Average distance from Sun: 482.5 million miles (778.3 million km)

Length of a year: 11.9 Earth years

Number of moons: 16

Saturn

Saturn is the second largest planet in the Solar System. It takes 29.5 Earth years to make a complete orbit of the Sun. It was not long before early astronomers began to realize that Saturn is the most beautiful planet in the Solar System. Even early telescopes revealed a spectacular ring system. Saturn's rings are made up of thousands of "ringlets" small bits of rock and ice, all held together in an orbit by the planet's gravity.

Saturn also has at least 23 moons and may have many more. Some moons are within the ring system and are called "shepherd" moons, because they appear to help to keep the ring system in place.

RINGS OF COLOR

Saturn's colorful rings do not actually touch the planet but are tilted at an angle to its orbit. The three main rings can be seen from Earth through a telescope – the outermost ring may be 200,000 miles (300,000 km) wide. Saturn's rings are not solid, and the light from bright stars shines straight through them.

Saturn is like a smaller version of Jupiter, its atmosphere consisting mainly of hydrogen. Under the clouds, which rotate around the planet every 10 hours, are thick lakes of liquid hydrogen. The rocky core is 12,000 miles (20,000 km)

Outermost ring visible from Earth

Brightest ring

A SYSTEM OF RINGS

Saturn's ring system is made up of billions of "snowballs" of rock and ice. They range in size from small flakes to chunks over 30 feet (9 m) wide.

TITANIC MOON

Saturn's largest and most famous moon is Titan, which has a diameter of 3,187 miles (5,140 km). Titan is the only moon in the Solar System with an atmosphere. Its thick atmosphere is made of nitrogen and methane gas. Titan may have lakes of liquid gas on its surface. Another of Saturn's moons is Mimas, with an enormous 80-mile (130-km)-wide crater on its surface. The crater looks like the giant eye of an alien.

SATURN FACT FILE

Diameter: 73.970 miles (119,300 km)

Average distance from Sun: 585 million miles (1.43 billion km)

Length of a year: 29.5 Earth years

Number of moons: 23

URANUS

Uranus was the first planet to be "discovered" — by English astronomer Sir William Herschel in 1781. It circles the Sun at an average distance from the Sun of 2 billion miles (2.87 billion km). The poles of this rather unusual planet point almost sideways, at an angle of 98 degrees. It is possible that Uranus was knocked over by a large body in the early history of the Solar System. Uranus is a bluish-green gas planet made up mainly of hydrogen and helium, with some methane. It has a very thin ring system of at least nine faint rings. The rings, which consist of rocks and dust, were only detected by astronomers in 1977. Uranus also has five large moons — Miranda, Ariel, Umbriel, Titania, and Oberon — as well as 10 small moonlets.

A BLUE-GREEN PLANET
The methane gas in the atmosphere of Uranus gives the planet its bluish-green color. Methane gas accounts for about one-seventh of the atmosphere. Passing spacecrafts found streaks of cloud in the planet's upper atmosphere. Astronomers know very little about the surface of Uranus.

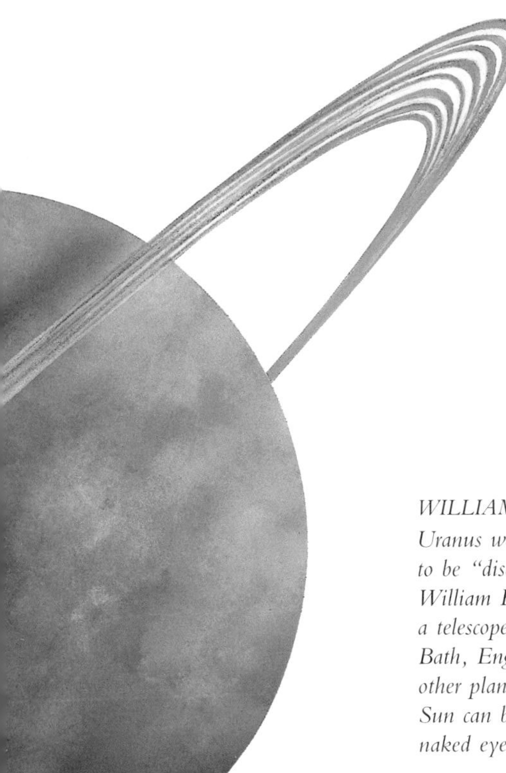

HERSCHEL'S DISCOVERY

Herschel made his discovery using a telescope in his garden in Bath, England. He compared the position of a star like object in the night sky against a star map and noticed that, during the following days, the object appeared to move in relation to the stars. He had discovered the seventh planet. Earlier astronomers had seen Uranus but had not noticed its slight movement in the sky.

WILLIAM HERSCHEL
Uranus was the first planet to be "discovered" __ by William Herschel using a telescope in 1781, in Bath, England. All the other planets closer to the Sun can be seen with the naked eye.

URANUS FACT FILE

Diameter: 31,500 miles (50,800 km)

Average distance from Sun: 1.75 billion miles (2.87 billion km)

Length of a year: 84 Earth years

Number of moons: 15

NEPTUNE AND PLUTO

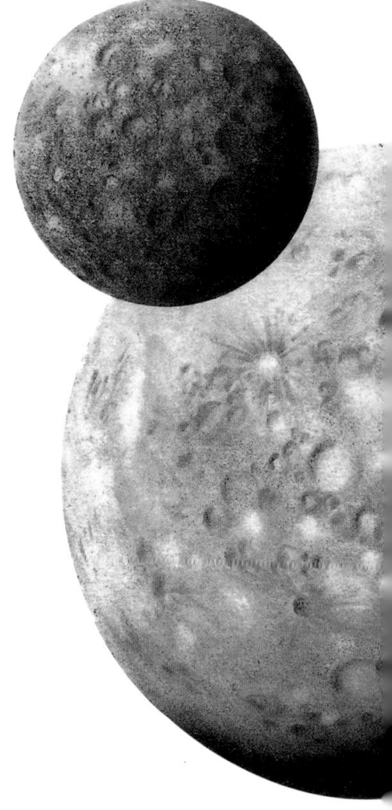

Neptune is a gaseous planet made of hydrogen and helium. It is very like Uranus but it is largely blue in color due to the composition of its atmosphere. Neptune has a very active atmosphere with high-speed winds that swirl around the planet faster than it rotates. The winds that carry "scooter clouds" around the planet at speeds of 1,500 miles per hour (2,400 km/h). Neptune's orbit sometimes extends beyond the orbit of Pluto, as it did from 1979 to 1999. Neptune has two large moons, Triton and Nereid. Triton, one of the largest moons in the Solar System, is unusual because it moves in a circular orbit from east to west.

Pluto may well have been a moon of Neptune at one time. Its surface probably consists of frozen water, ammonia, and methane. Pluto is the most distant planet from the Sun. It takes 248 years to orbit the Sun and won't return to the position it was discovered in until 2177! It was discovered in 1930 by an American astronomer, Clyde Tombaugh.

THE DOUBLE PLANET

Pluto was the last planet to be discovered, and even the most powerful telescopes on Earth only show it as a star like point of light. Pluto's moon, Charon, was discovered in 1978. It is almost the same size as Pluto, and the Pluto–Charon system is rather like a double planet. Charon circles Pluto every 6.3 days. Because Pluto rotates every 6.3 days, Charon appears to be stationary in the sky, like a geostationary satellite orbiting the Earth.

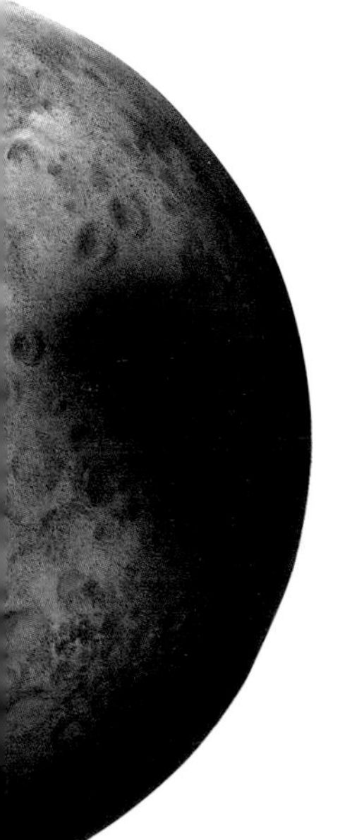

NEPTUNE FACT FILE

Diameter: 30,690 miles (49,500 km)

Average distance from Sun: 2.79 billion miles (4.5 billion km)

Length of a year: 165 Earth years

Number of moons: 8

PLUTO FACT FILE

Diameter: 1,417 miles (2,285 km)

Average distance from Sun: 3.65 billion miles (5.9 billion km)

Length of a year: 248 Earth years

Number of moons: 1

PLANET DISCOVERIES

Astronomers predicted the existence of another planet beyond Uranus before it was finally discovered by a German, Johann Galle, in 1846. They had noticed that Uranus was not staying in a completely stable orbit around the Sun, and suspected that it was being pulled slightly off course by the gravity of another body. Mathematicians calculated where this possible planet might be, and Neptune was found in almost the exact spot predicted by Galle.

Pluto was discovered by Clyde Tombaugh after an extensive search which began at the Lowell Observatory, Arizona, USA. He compared photographic plates of the night sky and saw that a 'star' had moved.

COMETS

Comets are also made from material left over when the Solar System was formed. They are like "dirty snowballs" of rock, dust, and ice. They travel in various orbits around the Sun, usually going deep into the far reaches of the Solar System. The orbits of some comets bring them close to the Sun after many years in darkness. When they come near to the Sun, comets reflect the Sun's light and can be seen in our sky. The Sun's heat and light also make comets shed material, which normally forms into the characteristic long tail.

One of the most famous comets is Halley's Comet, which appears in our skies every 76 years. When it last came close to the Sun, in 1986, it was rather a disappointing sight. Recently, a much more spectacular comet was Hale–Bopp. It shone brightly in the night skies in 1996–1997, and had a spectacular double tail.

COMET KOHOUTEK

Comets are named after the people who discovered them. Lubos Kohoutek found a comet in 1970. It was observed from space by the Skylab 4 *crew in early 1974.*

COMETS IN ORBIT

As a comet approaches the Sun, the heat makes it expand, evaporating gas and releasing dust. The gas and dust form a fuzzy head and a long tail. About 400 comets take between 3 and 200 years to orbit the Sun. There are about 500 known comets that will not return to the region around the Sun for thousands of years.

Tail of gas and dust

HALLEY'S COMET

The first recorded sighting of Halley's Comet was in 86 BC. After another appearance in 1301, the Italian artist Giotto di Bondone depicted it in his famous painting, *The Adoration of the Magi*. The comet appeared again in 1066, at the time of the Battle of Hastings. It can be seen on the famous Bayeux Tapestry. It was named for Edmond Halley who, in the early eighteenth century, realized that sightings in 1682, 1607, and 1531 must have been of the same comet. He predicted its appearance in 1756, and the comet was named for him.

METEORITES

Another kind of material left over from the formation of the Solar System consists of rocks of all shapes and sizes, and grain like particles of dust. These are meteoroids. They enter Earth's atmosphere at speeds of up to 30 miles (50 km) per second, burning up to leave behind a visible trail of hot gases called a meteor, or shooting star.

ARIZONA CRATER

A huge meteorite is thought to have hit the Earth in about 25,000 B.C. It created a huge crater 4,875 feet (1,265 m) wide and 574 feet (175 m) deep in Canyon Diablo, Arizona. The meteorite crashed with the force of a huge nuclear bomb.

METEORITE ROCK
Some small meteoroids survive the heat of entry into the Earth's atmosphere and have been recovered. It is possible to see about 10 meteors, or shooting stars, an hour on a clear night. During major periods of meteor showers, more than 100 meteors an hour may be seen.

Showers of meteors tend to occur during certain periods of the year. For example, the Earth encounters dust from Halley's Comet, which forms meteor showers in May and October. As these appear in the part of the sky where the constellations of Aquarius and Orion are at the time, they are called the Aquarids and the Orionids. Some of the larger rocks, which cause the occasional spectacular shooting star, survive the high-speed entry and reach the Earth. These are called meteorites. Several very large meteorites have hit the Earth in its history, forming craters that we can still see today.

METEORITE CLUES

About 500 meteorites hit the Earth each year. The largest known meteorite was found at Grootfontein in Namibia, southwest Africa, in 1920. It is 9 feet (2.75 m) long and 8 feet (2.43 m) wide. Recovered meteorites provide scientists with an opportunity to study some of the oldest original material in the Solar System. Grains of dust from a meteorite that fell in Murchison, Victoria, Australia on September 28, 1969 are thought to be older than the Solar System itself.

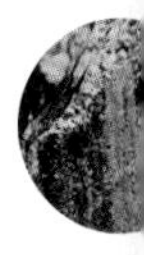

SPACE EXPLORATION

Spacecrafts have now visited every planet in our Solar System, except the most distant planet, Pluto. Our exploration of space began when the first artificial satellite, *Sputnik 1*, circled the Earth in 1957. Within a few years, both the United States and the Soviet Union were sending rockets to the Moon to prepare the way for the first astronauts to land on its surface. Spacecrafts have been sent to explore the planets, some flying past their target or orbiting around it, while

others land on the planet's surface. These spacecrafts send back valuable images and other data, allowing scientists to build up increasingly detailed information about our family of planets. Spacecrafts have investigated other objects in space too, including comets and the band of asteroids between Jupiter and

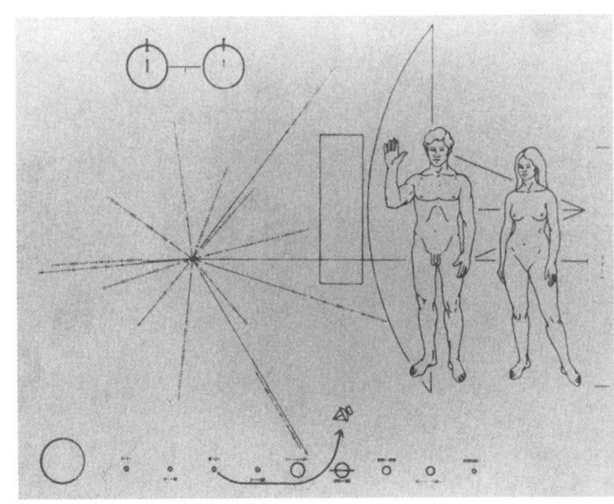

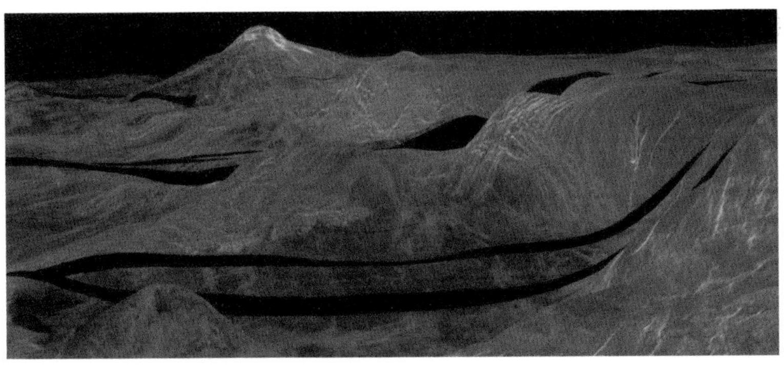

LIFTING OFF

For hundreds of years the only way to explore the planets was by using telescopes. All this changed in December 1962, when the American spacecraft *Mariner 2* flew past Venus. It sent back the first data about the planet, indicating that it was an extremely hot place. Since then, spacecrafts have explored every planet in our Solar System, except Pluto, and have also visited a comet and some asteroids.

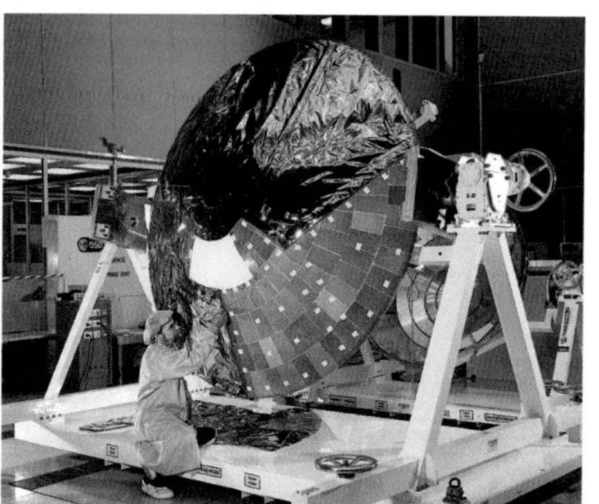

Craft have landed on Venus and Mars, and have penetrated the swirling atmosphere of Jupiter. Other spacecrafts have orbited Venus, Mars, and Jupiter, while several planets have received fleeting visits from passing spacecrafts. Closer to the Earth, the first rocket was launched to the Moon in 1958, and since then 12 men have walked on the Earth's nearest neighbor. Spacecrafts have also explored the Sun closely. Through space technology our knowledge of the Solar System has reached new and very exciting limits.

A VISITOR TO SATURN
The US/European Cassini–Huygens *spacecraft took 10 years to build. It is now on its way for a rendezvous with the planet Saturn in July 2004.* Cassini *will orbit Saturn while the* Huygens *probe will land on its moon, Titan.*

FIRST LANDING ON MARS
A Titan III-E Centaur D1 booster rocket launched the Viking 1 *spacecraft on August 20, 1975. The NASA spacecraft made the first soft landing on Mars.*

ANOTHER WORLD

The *Huygens* probe is expected to reveal Titan's surface as a world of methane seas, methane icebergs, methane snow, and a mixture of ice and rock. It will be dark, because the sunlight that filters through the orangy clouds gives a similar light to the light provided on Earth by a Full Moon. The main gas in Titan's atmosphere is nitrogen, and some scientists think that the orangy clouds may contain some organic material. This material is similar to that which has been created by scientists who are trying to simulate the formation of life on Earth.

LANDING ON TITAN
The European Space Agency's probe Huygens *should touch down on Saturn's moon, Titan, in November 2004. If this happens, it will be the first landing on the moon of a planet other than the Earth.*

EXPLORING THE SUN

The Sun has come under detailed scrutiny by a whole fleet of spacecraft since 1959, when the *Pioneer 4* spacecraft entered solar orbit after missing the Moon. Scientists were particularly interested in the particles of solar energy which affect the Earth and its upper atmosphere. They also wanted to try and understand how the Sun actually works. In 1962, the United States launched missions of Earth-orbiting solar observatories, called OSO. Later, two Helios spacecrafts and several other Pioneer spacecrafts were sent to operate in orbit around the Sun.

Special instruments on board the space station *Skylab* in 1973 and 1974 took images of the Sun in different wavelengths. They revealed activity within the Sun's atmosphere that cannot be seen in visible light. *Skylab* carried its own solar telescope mounted on the outside of the station. More recently, the European Solar and Heliospheric Observatory (SOHO) was launched to conduct non-stop observations of the Sun, rather like a solar weather station.

SOHO OBSERVATORY
The European Solar and Heliospheric Observatory (SOHO) was launched in 1995. It will conduct the most comprehensive observation so far of the Sun and its radiation.

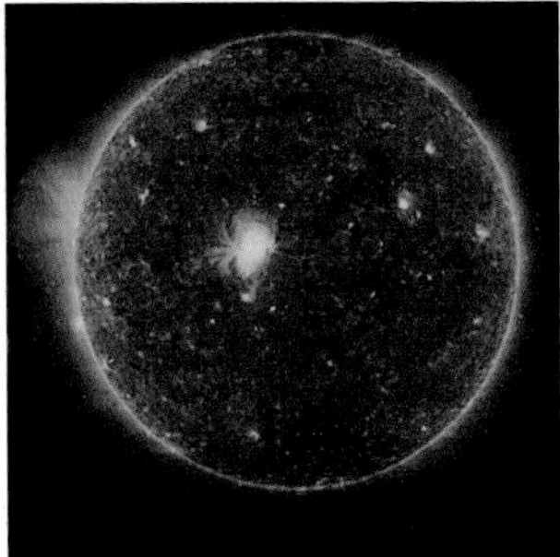

3.6 million degrees
Fahrenheit
(2 million degrees Celsius)

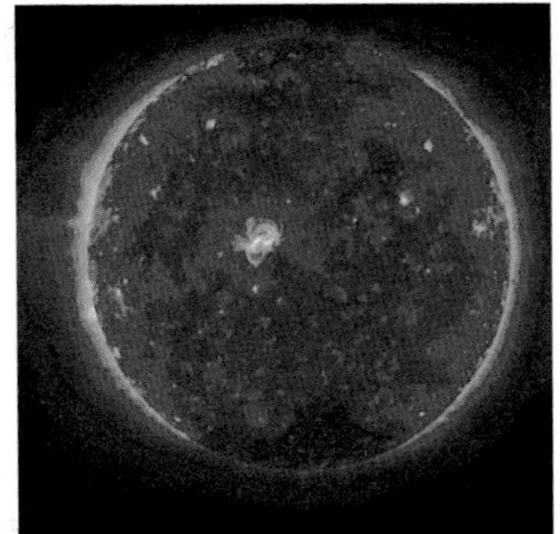

2.7 degrees million
Fahrenheit
(1.5 million degrees Celsius)

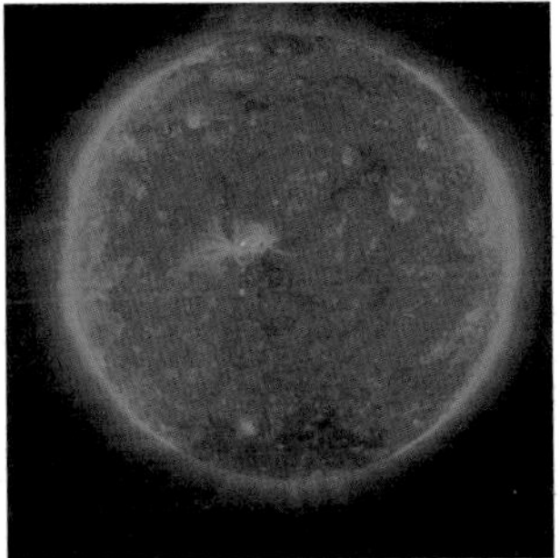

1.8 million degrees
Fahrenheit
(1 million degrees Celsius)

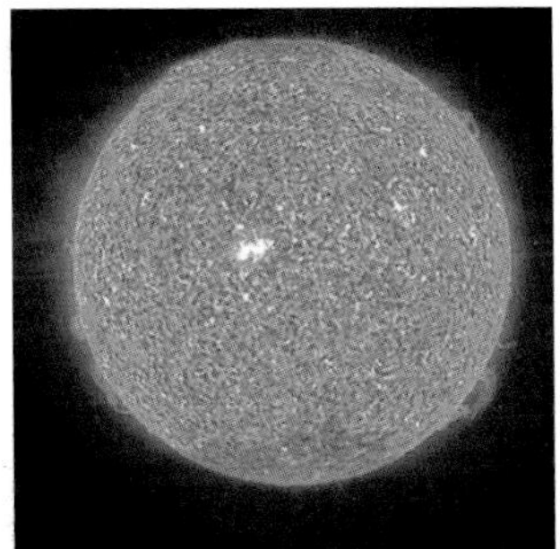

144,000 degrees
Fahrenheit
(80,000 degrees Celsius)

SURVEYING THE SUN

SOHO was placed into a special orbit 900 million miles (1.5 billion km) from the Earth. This orbit gives SOHO an uninterrupted view of the Sun from a point in space where the forces of gravity from the Sun and the Earth are equal. SOHO is part of an international program involving satellites from many countries. Their intensive survey of the Sun and its effect on the Earth is called the Solar Terrestrial Science Program.

IMAGES FROM SOHO
These ultraviolet images from SOHO show the different temperatures in the Sun's atmosphere. SOHO studied the Sun's surface activity, atmosphere, and radiation, including the solar wind. The study was made from a unique orbit that provided an uninterrupted view of the Sun.

VISITING MERCURY

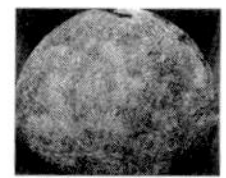

It is almost impossible to view Mercury's surface with a telescope. In fact, until 1973 almost the only thing we knew about the closest planet to the Sun was that it must be very hot! The first, and so far the only, spacecraft to visit Mercury is *Mariner 10,* which was launched on November 3, 1973. Its flight path took it to Mercury using a fly-by of Venus. The spacecraft used the gravitational force of Venus as a "sling shot" to divert it onto the right course.

Mariner 10 made three separate fly-bys of Mercury, coming to within 3,576 miles (5,768 km) of the planet on March 29, 1974, within 436 miles (703 km) on September 21, 1974, and within 29,803 miles (48,069 km) on March 16, 1975. The images from the spacecraft were astonishing. Mercury is just like the Moon! It has thousands of craters, including a huge meteorite crater called the Caloris Basin.

VIEW FROM MARINER 10
Mariner 10 *gave us the first and only clear view of Mercury, showing it to be surprisingly like our Moon.*

MARINER 10 MISSION

Mariner 10 was a cut-price spacecraft flying a cut-price mission! To save money it was launched by a less powerful booster rocket, and as a result it needed to use Venus as a "sling shot". This meant that *Mariner 10* had to pass within a 240-mile (400-km) target area about 3,000 miles (5,000 km) from Venus, or else it would miss the opportunity to visit Mercury. To do this, the spacecraft had to fire its own engine very accurately several times during its journey.

FLYING PAST MERCURY
Mariner 10 *was launched in 1975. It was the first spacecraft to be sent to explore two planets in a single mission. It flew past Venus once and past Mercury three times.*

VENUS UNVEILED

Venus is a hostile planet and has posed a great challenge to space engineers. They have nonetheless succeeded in mapping almost the whole planet and have even landed crafts on its surface. Venus is surrounded by thick clouds of carbon dioxide gas that create a surface temperature of about 830 degrees Fahrenheit (460 degrees C).

The planet has an atmospheric pressure 90 times that of the Earth. It was first explored successfully by *Mariner 2* in 1962. The Soviet craft *Venera 4* penetrated the clouds, sending back some data in 1967, and in 1970 *Venera 7*'s capsule reached the surface still intact. *Veneras 9* and *10* became the first Venus orbiters in 1975, and their landing capsules sent

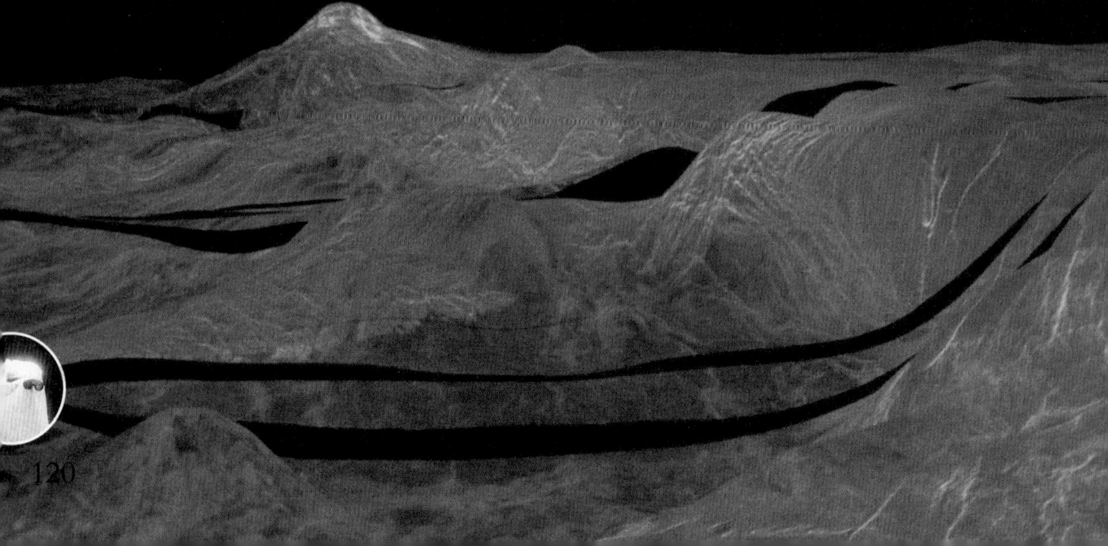

FLYING PAST VENUS
The first successful explorer of our Solar System was the 446-pound (203-kg) Mariner 2 *spacecraft. It flew past Venus at a distance of 21,080 miles (34,000 km) on December 14, 1962.*

back the first pictures of the surface — an extraordinary feat in view of the harsh conditions. Further Venera probes included those that sent back some radar images that penetrated the thick cloud. More recently, the *Magellan* Venus orbiter was deployed from the space shuttle in 1989.

IMAGES FROM MAGELLAN
The Magellan *radar mapping satellite generated false-color images of what the surface of Venus looked like beneath the thick clouds.*

LANDING ON VENUS

The earlier Venera probes proved that these spacecrafts did not need parachutes to complete their landing on Venus's surface because its atmosphere is so dense. *Veneras 9* and *10* landed at a speed of 26 feet per second. They were protected by an ingenious shock-absorbing system, rather like a lifebelt. The capsule sat in the center of a ring that was inflated with gas before the landing. An airbraking disk on top of the capsule also served as the radio transmitter.

THE EARTH FROM SPACE

The first astronauts to travel to the Moon were also the first to see the Earth as it might appear to explorers from another planet. On seeing the Earth as a tiny part of an enormous Universe, the *Apollo 8* astronaut James Lovell described our planet as a "grand oasis in the vastness of space". The *Apollo 8* crew's famous picture of the Earth as seen from space seemed to sum up Lovell's feelings.

After traveling into space to explore the Moon, the Apollo crew came back with something much more precious for the world's population — an appreciation of our tiny Earth as a fragile planet. As a result, there was an enormous increase in people's concern for the Earth's environment. People were also struck by the lack of real boundaries on the Earth's surface, unlike the view of the Earth that we see in a geography atlas.

THE RISING EARTH
In this picture taken during the Apollo 11 *mission the Earth seems to be rising above the surface of the Moon. The astronauts actually saw the Earth coming into sight around the side of the Moon.*

OUR BEAUTIFUL EARTH
For any astronaut, a view of the Earth from space is a captivating sight. Photos can never fully reveal the Earth's beauty and color.

A FAMOUS PHOTO

Many later Apollo crews took different pictures of the Earth from space, some of which show it as a fine crescent in the dark sky. The famous *Apollo 8* photo of the Earth rising was used on a US stamp. Each crew member — Frank Borman, James Lovell and Bill Anders — claims that he took the picture, jokingly repeating the claim whenever the three meet at public events. Anders was the chief photographer on board *Apollo 8* and he took the picture — but nor before he was "ordered" to by Borman and Lovell, after protesting that it wasn't on his photo schedule!

VIEW OF THE WORLD
This classic view of the Earth was taken by the Apollo 17 *crew on their way to the Moon in December 1972. It clearly shows the land masses of Africa and Saudi Arabia, with the continent of Antarctica below.*

PROBES TO THE MOON

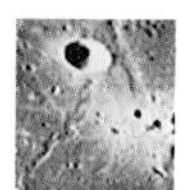

Soon after the first satellites were launched into orbit above the Earth, the next obvious target in space was the Moon. The first attempts to send probes to the Moon were made in 1958, but the first object to hit the Moon's surface was the Soviet *Luna 2* spacecraft in September 1959. Later, *Luna 3* flew round the far side of the Moon, revealing what it looked like for the first time. Ranger probes took close-up photos before they crashed onto the surface in 1964 and 1965.

The first soft landings on the Moon were made by Luna and Surveyor craft in 1966. The unmanned *Luna 16* brought back samples from the Moon in 1970, the year in which the Soviet Union also launched an unmanned lunar rover called Lunakhod. After the first astronauts landed on the Moon in 1969, unmanned flights became rarer. *Luna 24* made the final flight of this era in 1976.

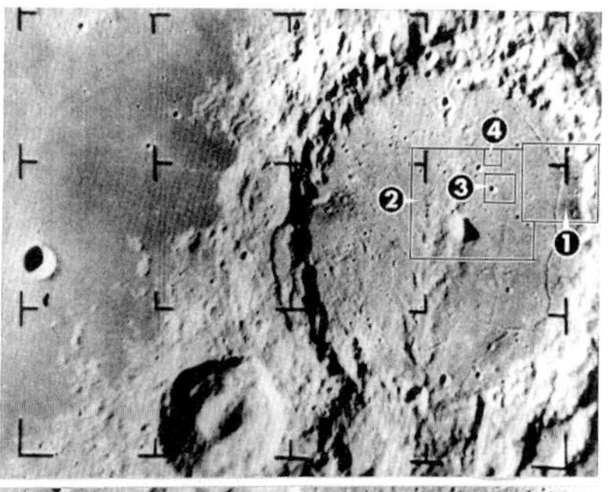

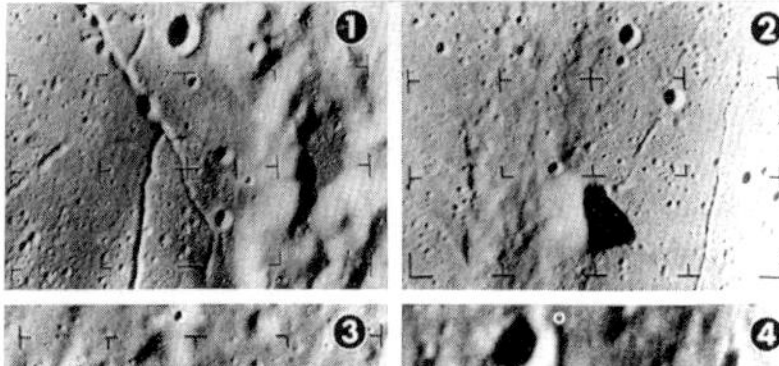

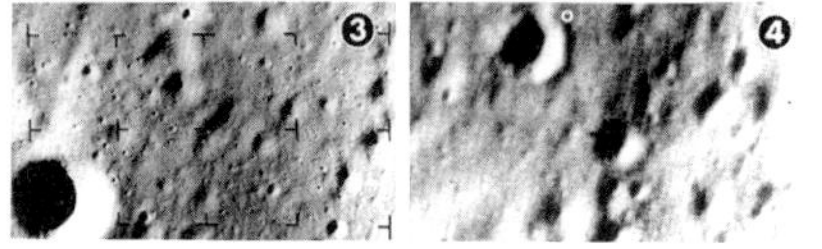

DESCENT TO THE MOON

These photos were taken on March 24, 1965 by the US spacecraft Ranger 9. It was *plunging toward the inside of the Alphonsus crater on the near side of the Moon.*

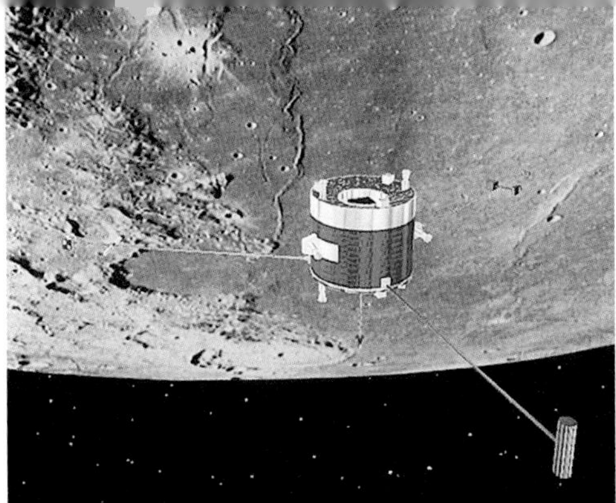

LUNAR PROSPECTOR
Launched in January 1998, Lunar Prospector *carried out a detailed chemical analysis of the Moon. The results confirmed that there might be frozen water under the bedrock in the polar regions.*

WATER ON THE MOON

New interest in unmanned exploration of the Moon began when the US spacecraft *Clementine* was launched in 1994. It sent back data indicating that there may be frozen water under the rock around the Moon's poles. In 1998, NASA's *Lunar Prospector* began orbiting the poles and seemed to confirm this information. The water exists as ice crystals in the soil. If Moon bases are built at some time in the future, a water-extraction plant on the Moon would be needed to make use of this resource.

SURVEYING THE MOON
The US Surveyor spacecraft made the first rocket-assisted touchdowns on the lunar surface in 1966 and 1967. Some of the Surveyor craft carried robot arms fitted with scoops to collect soil samples.

MISSIONS TO MARS

Mars has always held a particular fascination for the human mind because it is linked with the possibility that life may have existed on the planet at some time. The first Mars probe was launched in November 1960 but failed. This Soviet attempt was followed by many more, which, apart from *Mars 5* in 1974, all failed. In contrast to the failure of the Soviet missions to Mars, American spacecrafts to the planet met with spectacular successes. The first was *Mariner 4,* which took the first close-up images in 1965. *Mariner 9* went into orbit around Mars in 1971, and two Viking landers scooped up soil and took pictures in 1976.

In 1997 the *Pathfinder* spacecraft captured the world's imagination when it landed its *Sojourner* rover vehicle on the surface of Mars. The main quest now is to bring samples of Martian soil back to Earth. This task may be achieved by about 2007, after a series of lander–rover–orbiter and ascent vehicle missions that are due to start in 2003.

VIKING SPACECRAFT ON MARS
The first soft landings on the planet Mars were made by the American Viking 1 *and* 2 *spacecraft. They sent back the first photos of the Martian surface in 1976.*

IS THERE WATER ON MARS?

The Mars Climate Orbiter will enter polar orbit around Mars in September 1999. It will also act as a data relay satellite for the Mars Polar Lander, which is due to land on the edge of the largely frozen carbon dioxide polar cap about 600 miles (1,000 km) from the planet's south pole. A robotic arm will scoop up soil and deliver it to an on-board analyser. Scientists hope to find some evidence of the water that probably flowed across the Martian surface many years ago.

PATHFINDER TO MARS
After the Viking missions of 1976, the next soft landing on Mars was made 21 years later by the Pathfinder *spacecraft (far left). It carried a tiny roving vehicle called* Sojourner *(near left).*

LATEST VISITORS TO MARS
The next assault on the Red Planet got under way in December 1998 with the launch of the Mars Climate Orbiter *(top). This was followed by the* Mars Polar Lander *(bottom), which was launched in January 1999.*

FLYING PAST JUPITER

Four spacecrafts have explored Jupiter, the giant planet of the Solar System. The first was *Pioneer 10,* which was launched in March 1972 and flew past Jupiter at a distance of 80,600 miles (130,000 km) on December 5, 1973. One of its most spectacular images was a close-up of the Great Red Spot. *Pioneer 11* followed on December 3, 1974 at a closer distance of 26,000 miles (42,000 km). It used Jupiter's gravity to sling it onto a course to make a rendezvous with the planet Saturn. The next visitor to Jupiter was *Voyager 1,* which was launched in September 1997 and flew past the planet at a distance of 173,600 miles (280,000 km) on March 5, 1979. It was followed closely by *Voyager 2*, which was launched first but arrived on July 9, 1979, at the closest-ever distance of 400,000 miles (645,000 km). Jupiter's most recent visitor is the *Galileo* spacecraft (pages 130–131).

IMAGES FROM VOYAGER
Two Voyager spacecrafts flew past Jupiter in 1979. They returned spectacular pictures of the giant planet and its many moons.

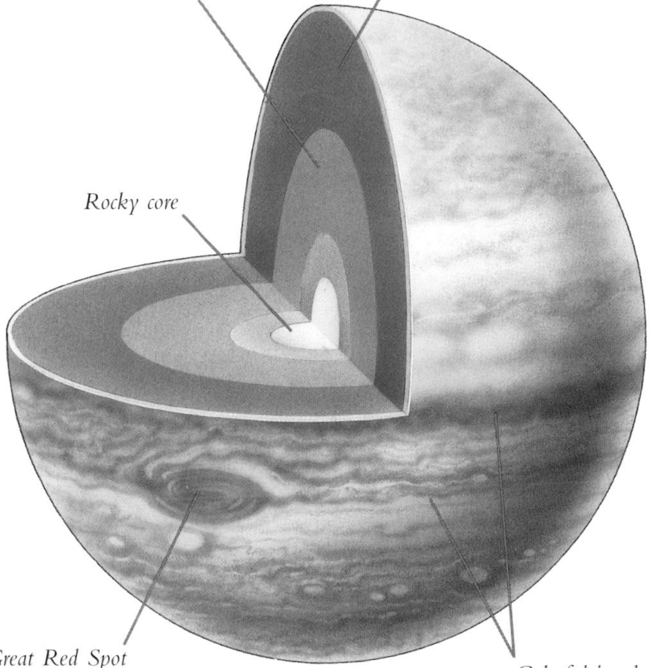

Liquid hydrogen

Metallic hydrogen

Rocky core

Great Red Spot

Colorful bands of clouds

INSIDE JUPITER

Most of Jupiter is made up of gases, mainly hydrogen and helium. The swirling clouds are divided into a series of bands of different colors, mostly white, brown, and orange. Inside the clouds are crystals of frozen ammonia and frozen water, and molecules of carbon, sulfur and phosphorus. Below the cloud level is a huge sea of liquid hydrogen, and then a layer of metallic hydrogen. Here the pressure on the hydrogen is so great that it starts to behave like a liquid metal. Electricity flows through the metallic hydrogen and creates a strong magnetic field around the planet. At the center of Jupiter is a solid core of rocky material — it is about 20 times more massive than the Earth.

BEYOND THE CLOUDS

Thick clouds swirl around Jupiter, hiding the planet's surface from view. The different bands of color in the clouds are known as belts (the dark bands) and zones (the light-colored areas between belts).

INTO THE UNKNOWN

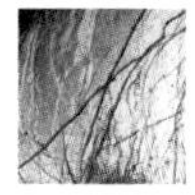

The space probe *Galileo* was deployed from the space shuttle in October 1989. It became the first Jupiter orbiter in December 1995. The probe entered Jupiter's atmosphere of thick cloud, plunging into the swirling unknown at a speed of 30 miles (47 km) per second. *Galileo* sent back data for a total of 57 minutes, reaching a distance of 98 miles (157 km) inside the clouds. It was finally defeated by very high temperatures and by an atmospheric pressure 22 times higher than the Earth's. Scientists were disappointed with the data from the probe. There seems to be just one layer of cloud, with wind speeds of 2,110 feet (643 m) per second caused by Jupiter's internal heat.

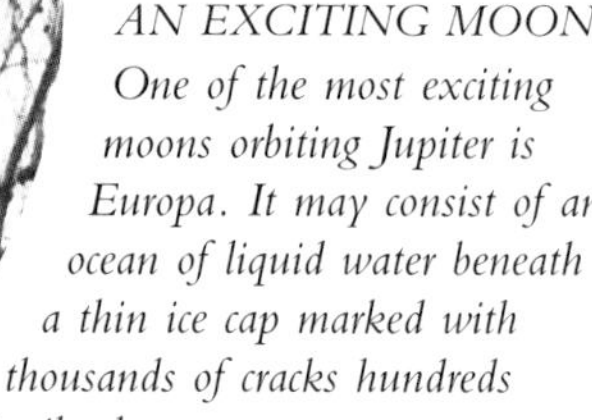

AN EXCITING MOON
One of the most exciting moons orbiting Jupiter is Europa. It may consist of an ocean of liquid water beneath a thin ice cap marked with thousands of cracks hundreds of miles long.

INTO JUPITER'S ATMOSPHERE
The first, and so far the only, craft to enter Jupiter's atmosphere is the Galileo *capsule. It plunged into the tops of the planet's dense clouds on December 7, 1995.*

THE GREAT RED SPOT

Jupiter's Great Red Spot was first noticed by an English astronomer, Robert Hooke, in 1664. The oval-shaped spot is 31,000 miles (50,000 km) long and about one-third as wide. It is big enough to swallow up four whole Earths. The Great Red Spot varies in intensity and color. For example, it has recently been measured as only 24,000 miles (40,000 km) long. It is a huge whirlpool of storm winds situated in the planet's southern hemisphere. The red color indicates that it contains a lot of phosphorus, which has been carried upward from the planet's interior. The Great Red Spot slowly changes its position from one year to the next. It can be observed through an ordinary telescope.

VISITORS TO SATURN

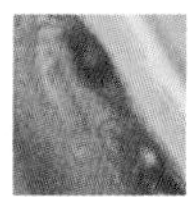

The first spacecraft to visit Saturn was *Pioneer 11*, which flew on to the ringed planet after its rendezvous with Jupiter. It passed the planet at a distance of 13,000 miles (21,000 km) on September 1, 1979. The first discovery it made was that Saturn's ring system does not consist of four divisions as seen in telescopes but of thousands of individual ringlets. Next, *Voyager 1*, which arrived on November 12, 1980 passing at a distance of 76,900 miles (124,000 km), followed by *Voyager 2* passing by at 62,600 miles (101,000 km) on 26 August 1981. The images returned from these crafts revealed Saturn's spectacular ring system in all its glory, as well as many of the planet's

CASSINI ORBITER
NASA's Cassini *spacecraft will deliver the* Huygens *probe into the atmosphere of Titan. It will then orbit Satrun and relay data from the probe back to Earth.*

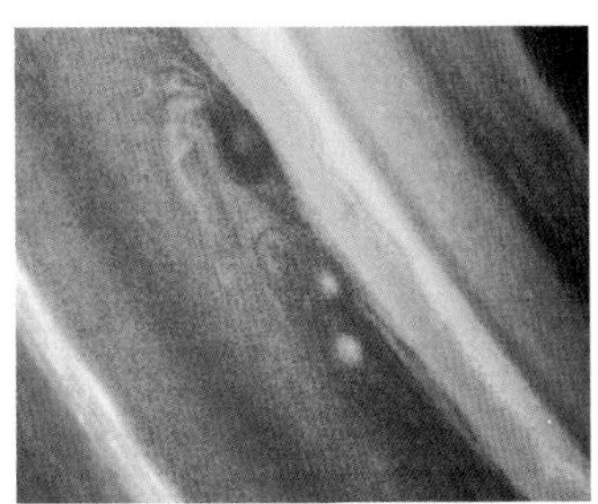

SATURN'S CLOUD
The dense layer of clouds that covers Saturn rotates around the planet every 10 hours. Under the clouds are thick lakes of liquid hydrogen surrounding an inner core of rocky material.

moons in great detail and its cloud bands. The next visitor to Saturn will arrive in 2004, when the US orbiter *Cassini* will become the first spacecraft to orbit the planet. It will send the European *Huygens* probe to land on the

RINGS AND RINGLETS

The Voyager spacecraft confirmed that Saturn has a seventh ring. The probes found that the planet's seven rings are really thousands of separate ringlets. Each ringlet consists of billions of objects ranging in size from icebergs 33 feet wide to tiny specks of ice smaller than a pinhead. The three small moons found in Saturn's ring system were named Prometheus, Pandora, and Atlas. They help to keep all the parts of the rings in place by means of small gravitational forces. These small moons were nicknamed the "shepherd" moons.

LANDING ON TITAN
The Cassini *spacecraft will carry the* Huygens *piggyback probe. The probe will parachute down onto the surface of Titan, Saturn's largest moon. Titan is one of the few moons in the Solar System to have an atmosphere.*

THE WORLD OF URANUS

Voyager 2 arrived for a close encounter with Uranus on January 24, 1986, at a distance of 44,000 miles (71,000 km). Until then, very little was known about the planet or its newly discovered ring system. The visit by the Voyager spacecraft changed all that. It found 10 new moons, all of them inside the orbits of the five known satellites. Two new rings were discovered, and two of the new moons seemed to be acting as "shepherds", rather like those discovered in the rings of Saturn.

Little was revealed of the planet itself beneath its greenish-blue atmosphere of hydrogen and helium. Because *Voyager 2* was being targeted

RINGS AROUND URANUS
This combined image was taken by the remarkable Voyager 2 *spacecraft. It shows part of the moon, Miranda, as well as the planet's ring system and Uranus itself.*

at another destination, Neptune, the craft's close flight path across Uranus lasted only about 5 hours. Signals from *Voyager 2*, which was 1.79 billion miles (2.88 billion km) away, took 2 hours 25 minutes to reach the Earth.

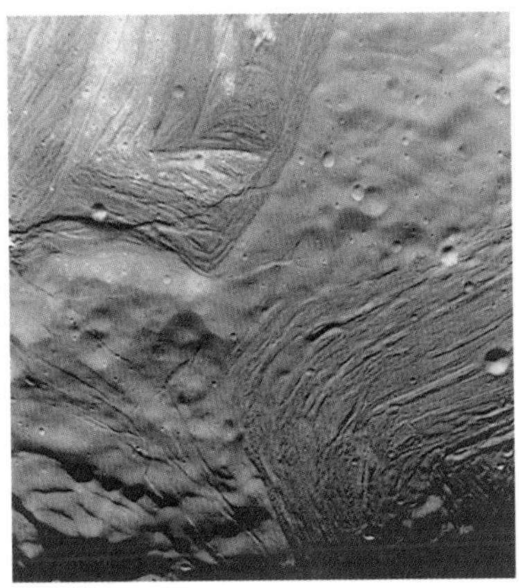

MIRANDA
Miranda is the smallest moon of Uranus. Its amazing surface is covered with a huge variety of features, including faults, grooves, terraces, and a steep cliff 10 miles (16 km) high.

FIVE MOONS

Voyager 2 made close-up images of the five known moons of Uranus: the cratered Oberon, bright Ariel, frosty Titania, dark Ombriel, and the amazing Miranda. Geologists have suggested that Miranda had fragmented at least a dozen times and re-formed in its present jumbled state, like a jigsaw puzzle put together in the wrong way. This theory may be linked to the fact that the rings of Uranus were found to consist of boulders rather than small particles. One of the ten new moons discovered, called Puck, was only 106 miles (170 km) across.

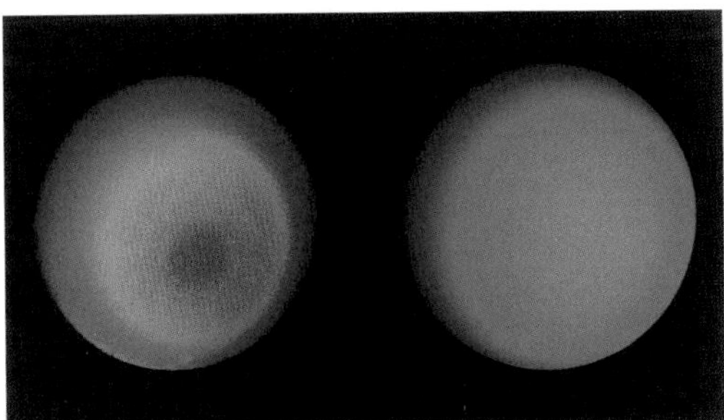

FALSE-COLOR VIEWS
Voyager 2*'s imaging system enabled it to take true and false-color views of Uranus. These highlighted the planet's atmospheric features and circulation patterns.*

JOURNEY TO NEPTUNE

Voyager 2 sped past Neptune at 17 miles (27 km) per second on August 24, 1989. Radio signals from the spacecraft took 4 hours, 6 minutes to reach the Earth 2.78 billion miles (4.48 billion km) away. The signals arrived with the power equal to a fraction of one billionth of a watt. However, they contained an incredible mixture of data showing a blue planet with three main features: a Great Dark Spot, a wide band of clouds and white "scooter" clouds. The Great Dark Spot, which is the size of the Earth, circles around Neptune's equator every 18.3 hours, compared with the planet's own 19-hour rotation. The 2,680-mile (4,320-km) -wide band of clouds in the southern hemisphere has a Lesser Dark Spot. The "scooter" clouds travel at a speed of 400 miles (640 km) per hour.

Neptune was proving to be a more turbulent planet than its neighbor

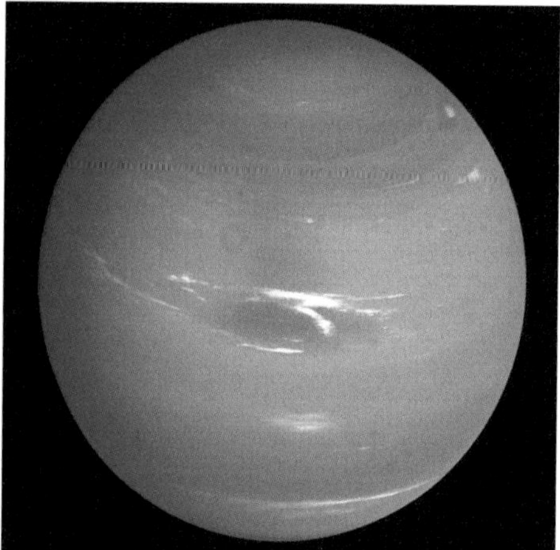

"SCOOTER" CLOUDS
The Great Dark Spot rotates counter-clockwise as it travels around Neptune. It is accompanied by white "scooter" clouds which are between 30 miles (50 km) and 60 miles (100 km) above the main atmosphere.

THE LESSER DARK SPOT
This view of Neptune's southern hemisphere shows the Lesser Dark Spot. It seems to "change lanes" during every circuit in the wide cloud bands.

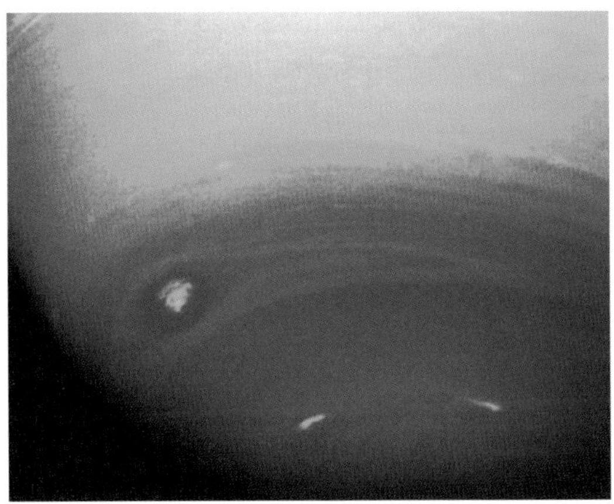

Uranus. The "scooter" clouds orbit mainly around the equator. They are like the cirrus clouds found on the Earth, but are made of methane ice. *Voyager 2* also found a ring system and four new moons, as well as imaging the remarkable world of Triton.

HALF AS BIG

The moon Triton, with a diameter of 1,860 miles (3,000 km), was only half the size that astronomers originally thought it was. Images of the moon revealed a surface of icy swamps of liquid nitrogen and methane, nitrogen frost clouds, mountains, craters and cliffs, quake faults, glaciers and geysers of liquid nitrogen. A surface temperature of −387 degrees F (−235 degrees C) makes Triton the coldest known place in the Solar System. Triton also displays glowing auroras caused by radiation trapped in Neptune's unique magnetic field. This radiation has also added a pinkish tinge to the moon's blue color.

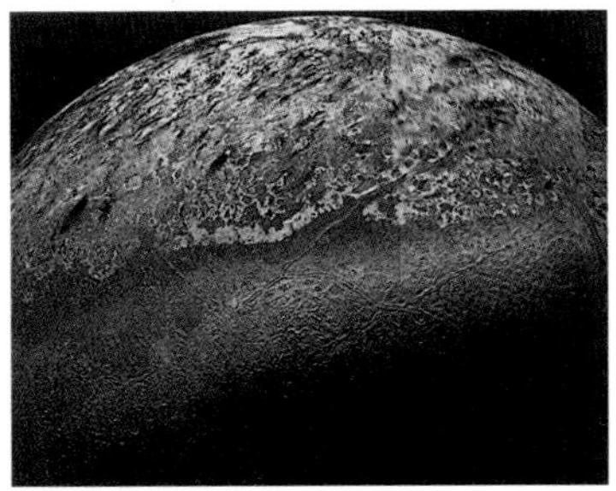

TRITON
Neptune's largest moon, Triton, is an extraordinary icy world of nitrogen and methane. Geysers on the moon's surface throw out liquid nitrogen 25 miles (40 km) up into space.

The Comet Explorers

As Halley's Comet approached the Sun in 1986 on its latest return journey from deep space, a fleet of international spacecrafts were approaching to meet it. They included the European *Giotto*, which was built in the UK.

Giotto flew an extraordinary mission, passing close to the comet's nucleus. The comet was shedding dust and gas in a stream of particles that battered against the craft's shield. The craft took close-up pictures of the nucleus, some of which showed small craters. *Giotto* survived to make a rendezvous with a second comet in 1992.

COMET CLOSE-UP
One of the best of the 2,112 images of Halley's Comet taken by Giotto *in very difficult photographic conditions shows a lumpy, irregular body about 9 miles (15 km) long and 6 miles (10 km) wide, with jets of dust and gas erupting into space.*

NASA is planning to launch a mission in its Discovery series in 2002. Called the Comet Nucleus Tour, or Contour, it will fly close to three comets between 2003 and 2008: Enke, Schwassmann–Wachmann 3, and d'Arrest. A possible later mission called Deep Impact will try to send a probe into the comet P/Tempel 1.

MEETING THE COMET

In 2002 Europe will launch the *Rosetta* spacecraft to rendezvous with the comet Wirtanen during its close fly-by of the Sun. *Rosetta* will make a long journey around the Solar System to meet the comet. The spacecraft will make a gravity-assisted flyby of both Mars and the Earth in 2005. It will pass within 300 miles (500 km) of the asteroid Mimistrobell in 2006 and will rendezvous with the asteroid Rodari in 2008 before finally reaching Wirtanen in 2011. It will travel with the comet on its solar flyby until 2013.

GIOTTO EXPLORES

Europe's British-built Giotto *spacecraft was the "star" of a fleet of crafts that explored Halley's Comet in March 1986.* Giotto *flew to within 360 miles (605 km) of the comet's nucleus.*

BEYOND OUR SYSTEM

Four spacecrafts — *Pioneer 10* and *11* and *Voyager 1* and *2* — are heading out of the Solar System and deeper into the Universe. At the end of 1998, *Pioneer 10* was about 6.55 billion miles (10.55 billion km) from the Earth traveling at a speed of 7 miles (12 km) per second. It is heading for the star Aldebaran 68 light-years away. It will take *Pioneer 10* 2 million years to reach the star. Contact with *Pioneer 11* has been lost. It is is heading toward the constellation Aquila and may pass one of its stars in 4 million years' time.

Voyager 1 is the most distant artificial object in space. It is 6.7 billion miles (10.8 billion km) from the Earth and heading toward an encounter with a dwarf star in the constellation Camelopardus in 400,000 years' time. *Voyager 2* is 5.2 billion (8.4 billion km) from the Earth, heading for a flyby of Sirius, the brightest star in the Earth's skies, in about 358,000 years' time. Engineers hope to keep communicating with *Voyager 2* until at least 2010.

THE PIONEER AGE

Pioneer 10, *which was launched in 1972 to visit Jupiter, is now heading toward the star Aldebaran in the constellation Taurus. The styles of the hair and clothing in this photograph show how much things have changed on Earth since the launch of* Pioneer 10 *almost 30 years ago.*

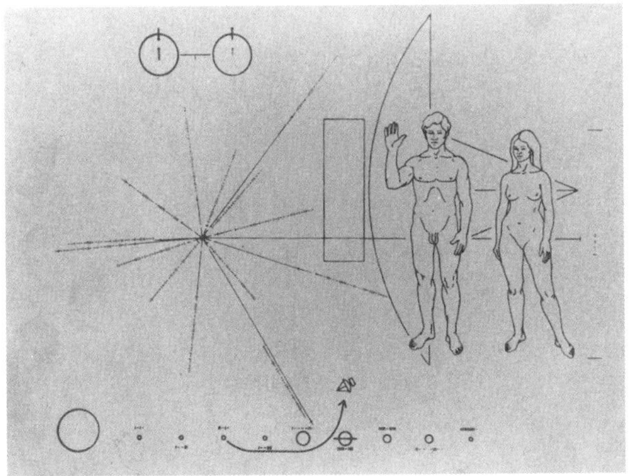

ON BOARD PIONEER

Pioneer 10 *carries a plaque that depicts a man and a woman and indicates the position of the planet Earth in space — just in case intelligent beings ever find the spacecraft.*

MESSAGE FROM EARTH

Each Pioneer spacecraft carries a gold-plated aluminum plaque measuring 6 by 9 inches (15 by 23 cm). The plaque is mounted on the spacecraft's antenna support struts, in a position that helps to shield it from erosion by interstellar dust. The plaque is etched with a diagram — it is a kind of picture message to any other intelligent beings that might exist in space. The diagram compares the height of the man and woman with that of the spacecraft. It also indicates the Earth's location in the Solar System in relation to the position of 14 pulsars.

TO THE STARS

Voyager 1 *is the most distant artificial object heading out of the Solar System. Traveling at a speed of 10.5 miles (17 km) per second, it is now heading toward the stars.*

ROCKETS AND SATELLITES

Rocket technology made great advances following the end of World War II, based initially on Germany's V-2 guided missile. During the 1940s both the United States and the Soviet Union were launching research rockets, and by 1957 the Soviet Union was able to use a rocket to launch the first satellite into space. The first US satellite, *Explorer 1*, was launched the following year.

Today's generation of powerful rockets are used as launch vehicles

for a whole range of different probes and satellites. Satellites in orbit around the Earth make it possible for us to have instant telephone communications, receive TV pictures from the other side of the world, and exchange information via computer. Weather satellites monitor the world's weather, enabling forecasters to predict weather extremes such as hurricanes more accurately.

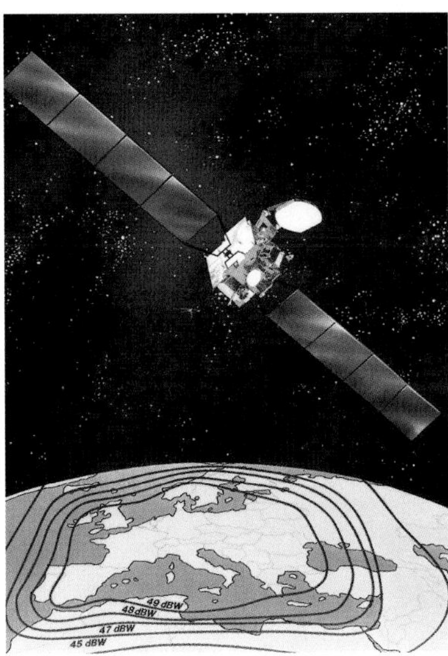

EARLY ROCKETS

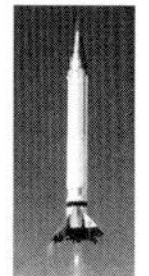

Rockets that used a solid propellant such as gunpowder were launched by the Chinese as long ago as A.D.1200. However, the first liquid-propelled rocket was not launched until 1926, making the first major breakthrough in the development of space travel. The Soviet Union, the United States, and Germany began developing these rockets in the 1930s and the German *V-2* was put to deadly use in World War II. At the end of the war, many German rocket engineers went to the United States and the Soviet Union, helping these countries to make more powerful boosters. The *V-2* was modified to fly with the upper stage of a US Corporal rocket. The Viking, a highly successful US rocket became the basis for one of the first satellite launchers. The *V-2/Corporal*, nicknamed "Bumper", the *Viking* and *Aerobee* rockets, and many Soviet rockets were used to carry scientific instruments and animals into the lower reaches of space.

THE V-2 *ROCKET*
The V-2 *rocket was developed by Germany during World War II. It fired over 3,000 missiles with deadly warheads. The rocket was later used by the USA to fly the first experiments into space.*

UP AND DOWN
Another early space rocket was the Viking. Like the V-2, *it flew an "up-and-down" flight, and not into orbit. It carried instruments that were later recovered after landing by parachute.*

VIKING SERIES

The Viking high-altitude research rocket was launched 12 times, starting in 1949. One flight, *Viking 4*, was made from a ship. Its flight was typical of the missions flown by these booster rockets fueled by liquid oxygen and kerosene. *Viking 4* was used to measure atmospheric density, the speed of upper atmosphere winds, and cosmic-ray emissions, and to photograph the Earth from space. In 1954 *Viking 11* reached a record height of 156 miles (252 km) .

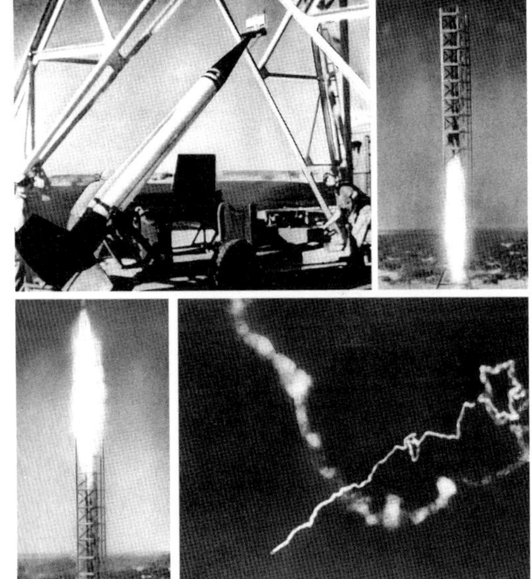

AEROBEE *ROCKET*
This Aerobee *rocket flew some of the first animals into space. The animals, such as mice, were recovered and examined to see how they had been affected by space travel.*

LAUNCHERS

By 1957, developing rocket technology had enabled the first Earth satellite, *Sputnik 1*, to be launched. Since then, all kinds of spacecraft have continued to provide a valuable service to the Earth. Science satellites help us to learn more about how the Earth is affected by solar winds and radiation. A fleet of weather satellites provides continuous monitoring of the whole Earth. Communications satellites span the globe providing television, radio, and telephone communications all over the world. The Internet could not exist without satellites.

Navigation satellites guide ships and aircrafts and even help to control the operation of commercial trucking. Environmental satellites survey the world's resources. Military satellites, including spy satellites, help to keep the peace. Space satellites help to run this age of high technology.

EUROPE'S ARIANE

The Ariane 5 *rocket will help Europe to continue to lead the commercial market for launching satellites into geostationary and other Earth orbits.*

EARTH WATCH
Many different kinds of satellites are continually monitoring the Earth, providing data about its changing environment and resources.

THE SPACE BUSINESS

Space is big business, and governments are not the only organizations involved in it. Private companies build satellites and the launchers that put them into orbit. They also supply the infrastructure to allow services to be provided by the satellites. The cost of a typical communications satellite is $250 million. To launch it on a vehicle such as *Ariane 5* costs an extra $100 million. Further money is spent on insurance and the launch itself. So even before a satellite starts to work, it has cost about $500 million to put it in orbit.

MONITORING HURRICANES
Weather satellites provide a vital service by detecting the birth and development of hurricanes. These satellites plot the path of a hurricane so that early warnings can be given to people living in the affected areas.

IN ORBIT

An object becomes a satellite when it is propelled fast enough — at a speed of 17,856 miles (28,800 km) per hour — to be able to fly in a continuous "fall", or arc, around the Earth without being pulled back by gravity. Satellites fly above the Earth's atmosphere at a height of at least 100 miles (160 km). The higher the orbit, the lower the speed of the satellite. A close-look observation satellite will fly in a low orbit around the Earth so that its cameras can take high-resolution images. However, a communications satellite is situated in a high orbit so that it can provide services to as large an area of the world as possible.

The angle, or inclination, at which a satellite orbits in relation to the Earth's Equator is important too. A

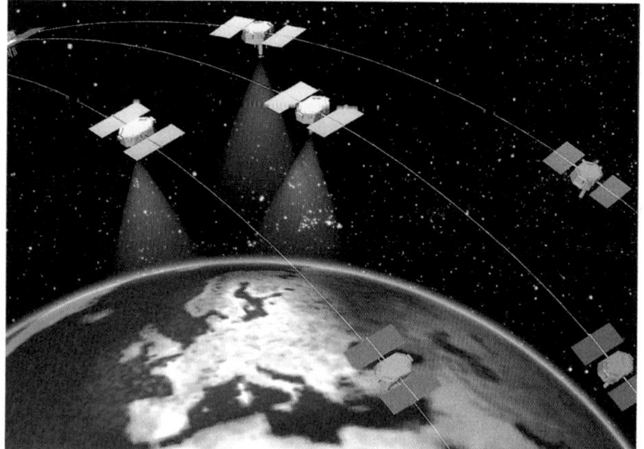

SATELLITE NETWORKS
Networks of satellites in low orbits around the Earth enable mobile phone users to make calls from almost anywhere in the world.

satellite in a low polar orbit will fly around the Earth every 90 minutes, completing 17 orbits in one day and crossing the Equator at an angle of 90 degrees. During this period the Earth will have rotated just once. In theory the satellite has covered every part of the world in a day.

GOING NOWHERE

Communications satellites fly in circular geostationary orbits round the Equator at a height of 22,890 miles (36,900 km). They travel around the Earth once in 23 hours, 56 minutes, 4.1 seconds, the same speed at which the Earth rotates. They therefore appear to be stationary in the sky. Three satellites positioned 90 degrees apart can provide coverage of the entire globe. The introduction of geostationary satellites resulted in a communications revolution.

GEOSTATIONARY ORBIT

Most communications satellites are in a geostationary orbit above the Equator. The satellite orbits the Earth in the same amount of time that the Earth takes to rotate once, so the satellite appears to stay in the same position in the sky.

SATELLITE TECHNOLOGY

Communications satellite technology has developed at a rapid rate. Just 30 years ago the rather basic Telstar satellite was transmitting TV signals to a ground station for onward transmission by land line to people's homes. Today, satellites can beam TV pictures directly into millions of homes — and from not just one channel but hundreds of different ones at the same time. Some important scientific advances have made this possible. First, we can now produce miniature components, such as traveling wave tube amplifiers. Also the satellites, with their perfectly fashioned antennas, have extremely high transmitting power — they are supplied with large amounts of electricity by advanced solar panels. Lastly, the high power transmissions

SIGNALS FROM SPACE
The antennas of communications satellites are "aimed" at specific user areas, called footprints. In these areas, highly amplified high-power signals are received by "dishes" that are as small as dinner plates.

mean that satellite receiving dishes can be smaller. Satellites provide numerous communications services other than television. Some are able to handle 30,000 simultaneous phone, data, and fax calls.

SATELLITE SYSTEMS
The Iridium satellite system uses 66 operational satellites in a low orbit above the Earth. The satellites are positioned in such a way that a mobile phone user is within "line of sight" of at least one satellite at any time.

SATELLITE POWER

The 155-pound (77-kg) Telstar satellite, with solar cells providing 15 watts of electricity, used to transmit to a single receiving dish with a diameter of 82 feet (25 m). Today, a typical communications satellite weighs 7,700 pounds (3,500 kg) and has 30 amplifiers providing high power transmission in many wavebands. Using 6,000 watts of power from solar cells, it transmits to thousands of receiving dishes as small as 3 feet (90 cm) in diameter.

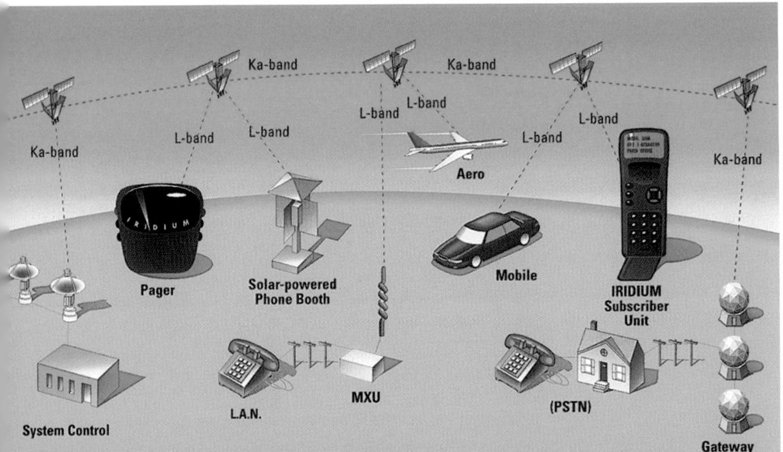

COMMUNICATING ON THE MOVE
Modern communications technology allows us to keep in touch with each other from almost anywhere in the world. The network of satellites in space provides instant communications whether by telephone, mobile phone, fax, pager, computer, or e-mail.

WATCHING THE WEATHER

A fleet of polar and geostationary orbiting satellites from the United States, Russia, Europe, India, and Japan provide a daily World Weather Watch service to the world's population. The satellites help meteorologists to make instant weather forecasts, including early warnings of weather extremes such as hurricanes. These early warnings enable thousands of people to evacuate danger areas, with the result that many lives are saved. The familiar weather satellite pictures we see on our TV screens are only one part of the weather story. Satellites also take images in different wavelengths, highlighting other aspects of the weather, such as temperature and atmospheric content.

Daily maps from satellites show the surface temperature anywhere on the Earth, both on land and at sea. Other images reveal the water content in the atmosphere and its pattern of circulation, and monitor the damage to the Earth's ozone layer caused by the emission of greenhouse gases.

TIROS WEATHER SATELLITE
The first weather satellite was the American TIROS 1, *launched in April 1960. TIROS images proved the ability of satellites to provide early warnings about hurricanes.*

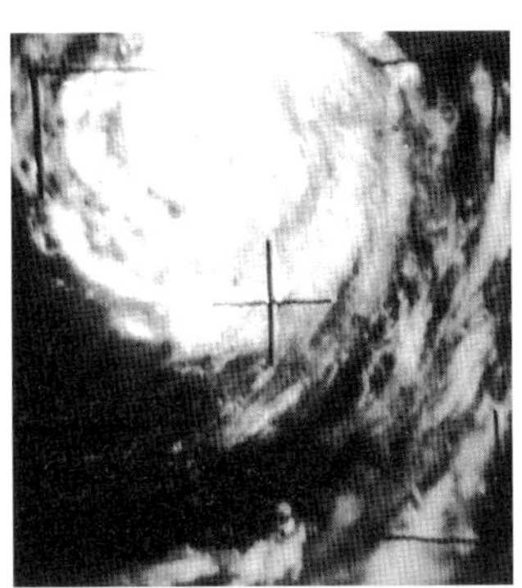

WEATHER PICTURES
The images from the TIROS 1 *satellite were rather crude. The cameras on this satellite always point to the same part of the Earth, so as the Earth spins round they can only photograph it for part of each day. Today's more advanced weather satellites point their cameras towards the Earth all the time.*

ENVIRONMENT WATCH

Meteosat's successor, called Metop, is a typical example of today's environmental satellite. It will be launched in 2002 and will carry nine instruments: a radiometer to monitor sea surface temperature, as well as the extent of ice, snow, and vegetation cover; a sounder to return data on the moisture content and height of clouds; a microwave sounder to measure humidity; and a scatterometer to measure sea surface winds.

WEATHER MONITOR
International meteorological satellites in polar and geostationary orbit above the Earth are continually monitoring the world's weather as well as changes to our environment.

NAVIGATION SATELLITES

Navigation satellites can provide accurate information to within a few arm's lengths of a person's location anywhere in the world — on land, at sea, or in the air — or in space. They can also work out the speed of a moving person or object to within 4 inches (10 cm) per second. This technology is vital for all types of military operations, from guiding a missile to its target or telling an undercover agent exactly where he or she is.

Russia and the United States operate fleets of navigation satellites. The US Air Force operates a navigation service called the Global Positioning System, or GPS. It consists of 24 Navstar satellites that are at all times equally spaced apart in six different paths around the Earth.

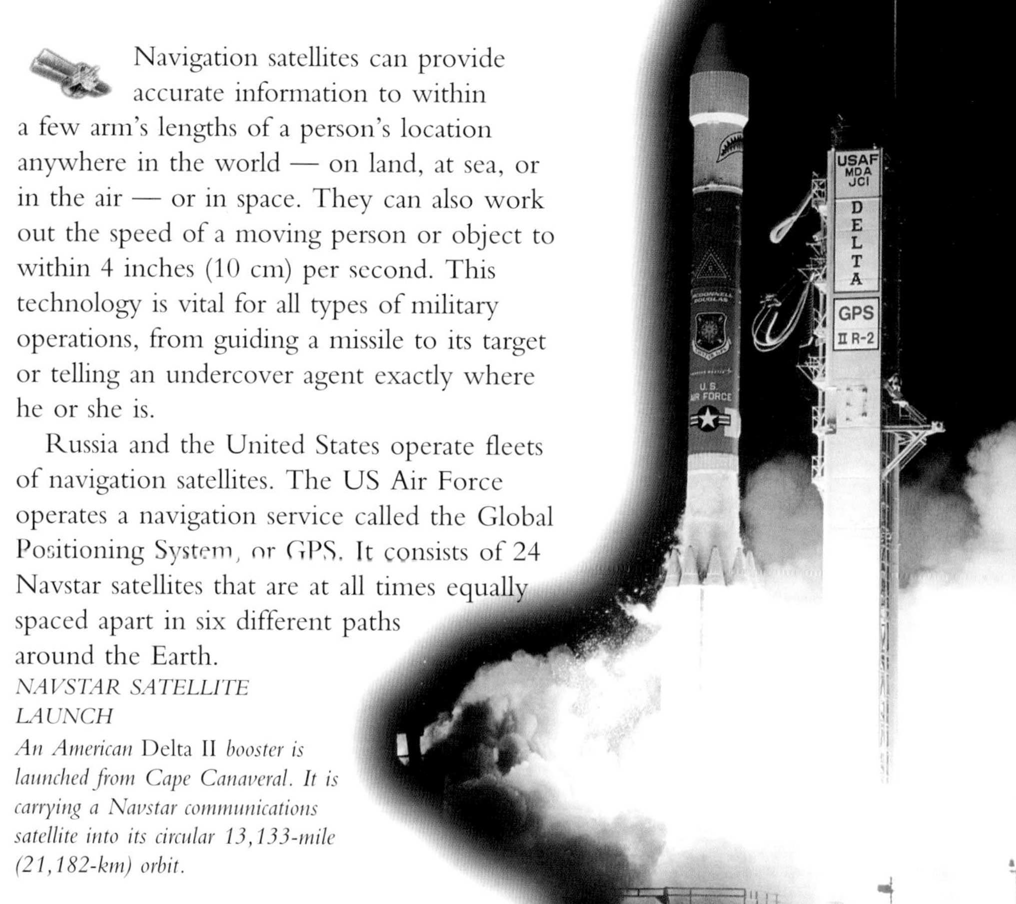

NAVSTAR SATELLITE LAUNCH

An American Delta II *booster is launched from Cape Canaveral. It is carrying a Navstar communications satellite into its circular 13,133-mile (21,182-km) orbit.*

NAVSTAR
Navstar satellites like this one provide vital positioning information for all kinds of uses, from guiding a military fighter through the skies to helping lost sailors find their way at sea.

HOW GPS WORKS

Each GPS satellite continually transmits its position and the exact time of its transmission. A receiver on an aircraft or ship or carried by a hiker, for example, receives signals from four satellites at the same time. The receiver processes the data and displays the user's position and speed and the exact time. GPS has become a vital part of worldwide search and rescue. A search beacon on a liferaft in the middle of the ocean can be located and positioned accurately by GPS satellites.

TRAFFIC CONTROL
Navigation technology that was originally developed for the military has now been transferred into the civilian sector. Soon even airlines will rely on satellites for air traffic control.

SCIENTIFIC SATELLITES

Many satellites have been launched for purely scientific purposes — to learn as much as possible about the Earth and its place in space. The discovery of radiation belts around the Earth by America's first satellite, *Explorer 1,* is a classic example of this. Other satellites have been launched to study the radiation belts, the Earth's magnetic field, and its upper atmosphere.

Many research and development satellites have been launched to test out new equipment to be used on future operational spacecrafts, while other satellites have conducted experiments using the lack of gravity in space, called microgravity. Some industrial processes, such as separation and mixing, can be improved greatly in microgravity. Space processing could lead to a revolution in the world's pharmaceuticals and electronics industries.

EXPERIMENTS IN SPACE

The European free-flying spacecraft Eureca *was equipped to carry out a range of experiments. It was deployed by the space shuttle in 1992 and left in orbit until its retrieval by another shuttle a year later.*

INTO THE UPPER ATMOPSHERE
The Upper Atmosphere Research Satellite was launched on board the space shuttle in 1991. It is studying in detail the mechanisms controlling changes in the upper atmosphere.

EURECA!

The name "Eureca" stands for European Retrievable Platform. It carried a total of 15 experiments involving materials processing, life sciences, radiation technology, and astronomy. Items on board included shrimp eggs, spores, solar monitors, and cosmic dust collectors. The Shuttle Pallet Satellite, or SPAS, is often deployed and retrieved during a shuttle mission. It carries experiments to evaluate the effects of microgravity — without any disturbance such as the vibrations that can be caused by the shuttle itself.

EXPLORER 15
Explorer 15 *was launched in October 1962 with the extraordinary objective of studying an artificial radiation belt. The belt had been created by a US high-altitude nuclear explosion, 250 miles (400 km) up in space.*

LAUNCH VEHICLES

As the space age has developed, communications satellites have become a major part of modern technology. As a result, several private operators have been established to provide communications satellite services, and a commercial launch vehicle market evolved to cater for the demand. It is now big business, with the majority of satellites needing to be launched in geostationary transfer orbit, a staging post for flights to geostationary orbit.

Launch providers can charge a customer up to $120 million for a launch, and many providers compete for business. The world leader is the European consortium Arianespace, which operates the *Ariane 4* and *5* rockets. International Launch Services offer the American Atlas and Russian Proton launch vehicles.

ATLAS LAUNCHERS
The first stage of this Atlas launch vehicle is based on the first US intercontinental ballistic missile. The commercial fleet of Atlas boosters is operated by International Launch Services.

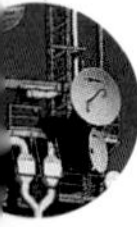

RUSSIAN BOOSTERS
The Russian Proton booster has been used to launch probes to the planets and modules for the Mir *space staton. It is also now carrying satellites for commercial customers.*

GETTING INTO ORBIT

Getting into geostationary orbit is a complicated affair. First, the launch vehicle usually places the satellite into a highly elliptical orbit with a low point of about 120 miles (200 km) and a high point of about 22,300 miles (36,000 km), The satellite is at an angle over the Earth's Equator. Then the satellite fires its own engine until its orbit gradually becomes circular with a constant height of 22,300 miles (36,000 km) above the Earth. The satellite is located directly over the Equator in a perfect geostationary orbit.

CHINA'S LONG MARCH
China is also in the commercial launcher business. It operates a fleet of Long March boosters, including the powerful Long March 3B.

LAUNCH SITES

The first satellite launch site was the former Soviet Union's Baikonur Cosmodrome in Tyuratam, Kazakhstan. *Sputnik 1* was launched from here in 1957. Baikonur is still Russia's main launching base, although the Plesetsk Cosmodrome inside Russian territory is also used. The main US launch site at Cape Canaveral, in Florida, is complemented by the adjoining space shuttle base at the Kennedy Space Center. Another major American launch site is at Vandenberg AFB, in California. It is primarily used for launches into polar orbits.

Europe's launch base is at Kourou, in South America, while China operates three sites, the main one being Xichang. Israel has launched satellites westward over the Mediterranean Sea from Palmachin. India's satellites are launched from Shriharokota. The US company Boeing is building a transportable launch platform to operate on the Equator in the Pacific Ocean.

CAPE CANAVERAL

The name of Cape Canaveral is synonymous with the space age. It was the site of the first US satellite launch. Further north is the Kennedy Space Center, where the first manned flight to the Moon was launched.

LAUNCH FROM JAPAN
Japan operates two satellite launch sites — at Tanegashima (above) and at Kagoshima. Launches of the large H2 satellite launcher take place from Tanegashima.

CAPE CANAVERAL

Cape Canaveral is a large sand spit jutting out into the Atlantic Ocean from the east coast of Florida. It used to be inhabited by American Indians and was probably seen by Juan de Leon, one of the first Europeans to discover America in 1509. In 1868 the small town of Cocoa was established south of Cape Canaveral and a lighthouse was built nearby. Today the lighthouse is often mistaken for a rocket by visitors to the resort of Cocoa Beach.

EUROPE'S LAUNCH SITE
The Arianespace commercial launch operation is based at Kourou in French Guiana, South America. Its location close to the Equator makes it an ideal place to start the journey of geostationary transfer orbit.

SPACE PROGRAMS

After the success of the first satellites in space, both the United States and Soviet Union embarked on their own space programs. The main part of each program consisted of a series of manned space flights. The Soviet Union's powerful rockets allowed them to launch large well-equipped spacecrafts able to sustain cosmonauts in space for several days. Meanwhile, the United States led the way with satellite technology

and also in the race to the Moon, which they eventually won.

The first flight of the Apollo Moon program took place in October 1968. In December 1968, the three astronauts on board *Apollo 8* orbited the Moon 10 times. By July of the following year, the first person had set foot on the surface of the Moon. American astronauts made five further landings on the Moon between 1969 and 1972.

RACE TO THE MOON

In 1961 President John F. Kennedy decided to respond to the Soviet "threat" of dominance in space. He announced to the US Congress on May25, that he wanted his country to land men on the Moon before 1970. The space project would cost more than $25 billion and would require a series of manned space flights to prove the technology, and Moon scouts to check out the new territory. What became known as Project Apollo was one of the most extraordinary undertakings of the twentieth century. It was undertaken on the assumption that the Soviet Union was also planning to send men to the Moon, so the "space race" became known as the "Moon race."

Space flight dominated the 1960s, hardly ever leaving the front pages of newspapers as important steps in the Apollo project were taken year by year. These events led up to the momentous first manned landing on the Moon in July 1969.

FIRST AMERICAN IN SPACE
Twenty-three days after Soviet cosmonaut Yuri Gagarin became the first person in space, Alan Shepard became the first American in space in the Mercury capsule Freedom 7.

SHEPARD'S SPLASHDOWN
US astronaut Alan Shepard is hauled aboard a helicopter after completing his 15 minute up-and-down space flight with a planned splashdown in the Atlantic Ocean.

A NATION ON THE MOON

President Kennedy said: "I believe that this nation should commit itself to achieving the goal, before this decade is out, of landing a man on the Moon and returning him safely to Earth. No single space project in this period will be more exciting, or more expensive to mankind, or more important for the long-range exploration of space; and none will be so difficult or expensive to accomplish. ... It will not be one man going to the Moon ... it will be an entire nation.

MAN ON THE MOON
President Kennedy's goal was to place an American on the Moon by 1969. When he announced the Moon project, the USA had just 15 minutes of manned space flight experience.

GEMINI AND SOYUZ

GEMINI RENDEZVOUS

One of the main successes on the road to the Moon was the first rendezvous mission in space. It took place in 1965 when the Gemini 6 *spacecraft met up with its sister ship* Gemini 7*, which was already in orbit.*

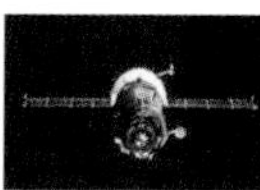

The US Gemini and the Soviet Union's Soyuz spacecrafts were built with Moon flights in mind. Gemini was to carry out most of the tasks required for the Apollo mission while remaining in orbit around the Earth. These tasks included rendezvous and docking and space walks. Soyuz was to be part of the manned lunar spacecraft as well as a manned lunar fly-by spacecraft. Soyuz eventually became a highly successful solo Earth orbiter and a space station ferry vehicle and is still used today.

The maneuverable Gemini two-man craft made 10 manned missions between 1965 and 1966, including four dockings in Earth orbit and space walks lasting over 2 hours. Soyuz was first launched in 1967 but failed, killing the lone cosmonaut on board.

Later Soyuz spacecraft docked together in space, demonstrated crew transfers and flew solo science research missions. They mostly carried crews to and from space stations.

"HELLO, SPACEMAN"

A rendezvous in space is a major feat in space exploration. It is not just a matter of flying up into space and meeting another vehicle. Complex orbital mechanics mean that a spacecraft has to make a series of vital engine firings to move into a perfectly matching orbit with the second vehicle, at exactly the same height, speed, position, and time as the other craft. For the *Gemini 6* and 7 rendezvous to be successful, it required seven carefully planned orbital maneuvers.

GEMINI LAUNCH

The two-man Gemini spacecraft is seen here being launched on a Titan 2 *intercontinental ballistic missile from launch pad number 19 at Cape Canaveral in Florida.*

SOYUZ SPACECRAFT

Different versions of the Soviet Soyuz vehicle were developed for planned lunar missions, for docking with space stations in Earth's orbit, and for solo flights.

First Apollo Flights

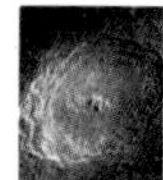

After the three-man crew of *Apollo 7* successfully tested the command and service modules in Earth orbit, a decision was made to send *Apollo 8* to the Moon. The Unite States feared that the Soviet Union was about to send two cosmonauts around the Moon on a fly-by mission. So the planned *Apollo 8* lunar module test flight in Earth orbit was canceled. Instead, *Apollo 8* was sent to make 10 orbits of the Moon. The mission was one of the biggest milestones in space history and one of the major events of the twentieth century.

The three astronauts in *Apollo 8* were launched on December 21, 1968. While orbiting the Moon over Christmas they sent back messages

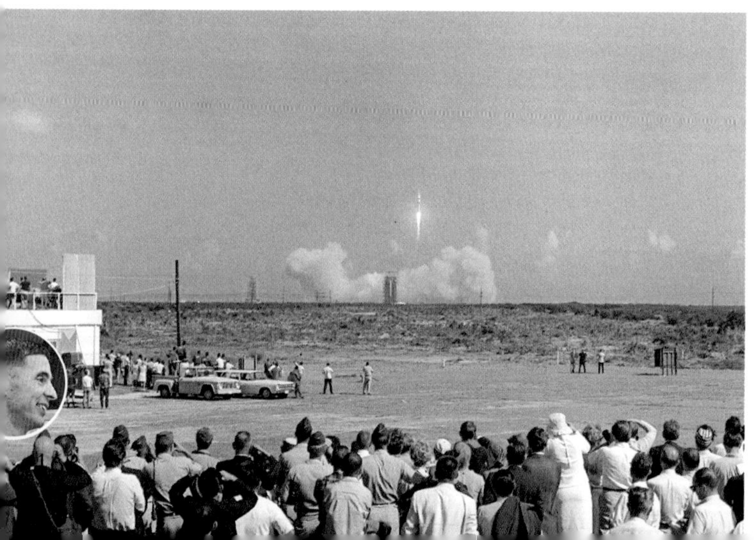

LAUNCH OF APOLLO 7
A smaller Saturn 1B *rocket boosts* Apollo 7 *into orbit on the first manned flight of the Apollo space program. The three-man crew made a successful 11-day flight, proving the command and service module systems.*

of goodwill and read from the Bible. *Apollo 8* came to within 66 miles (110 km) of the Moon's surface. The memorable flight ended with a safe splashdown in the Pacific Ocean — and the astronauts brought back the first photo of the rising Earth as seen from the Moon.

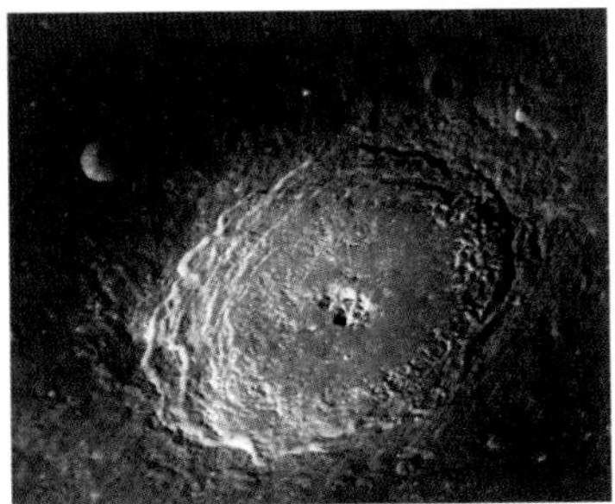

MOON CRATER
The first close look at the Moon was made during the Apollo 8 *mission. This is the crater Langrenus taken by* Apollo 8*'s photographer, Bill Anders, while orbiting the Moon.*

MOON QUOTES

Many memorable quotes were made by the *Apollo 8* crew. James Lovell gave the first description of the Moon: "the Moon is essentially gray, no color. Looks like plaster of Paris. Sort of grayish sand." He later remarked: "The vast loneliness of the Moon up here is awe-inspiring and it makes you realize just what you have back on Earth. The Earth from here is a grand oasis in the vastness of space." Frank Borman said that the Moon "is a vast, lonely forbidding type of existence, a great expanse of nothing."

CREW OF APOLLO 8
The first men to fly to the Moon were Frank Borman, James Lovell, and Bill Anders. They made 10 orbits of the Moon in their Apollo 8 *spacecraft during Christmas 1968.*

MAN ON THE MOON

Eight years after President Kennedy's pledge to land an American on the Moon, *Apollo 11* was ready to do just that. Millions of people all over the world followed one of the most historic events in the history of the human race — the first steps on another body in space. The actual landing was a nail-biting affair as several computer alarms threatened to end it. Flight commander Neil Armstrong had to take manual control to stop the lunar module landing in a rocky crater. He landed with just seconds of fuel left.

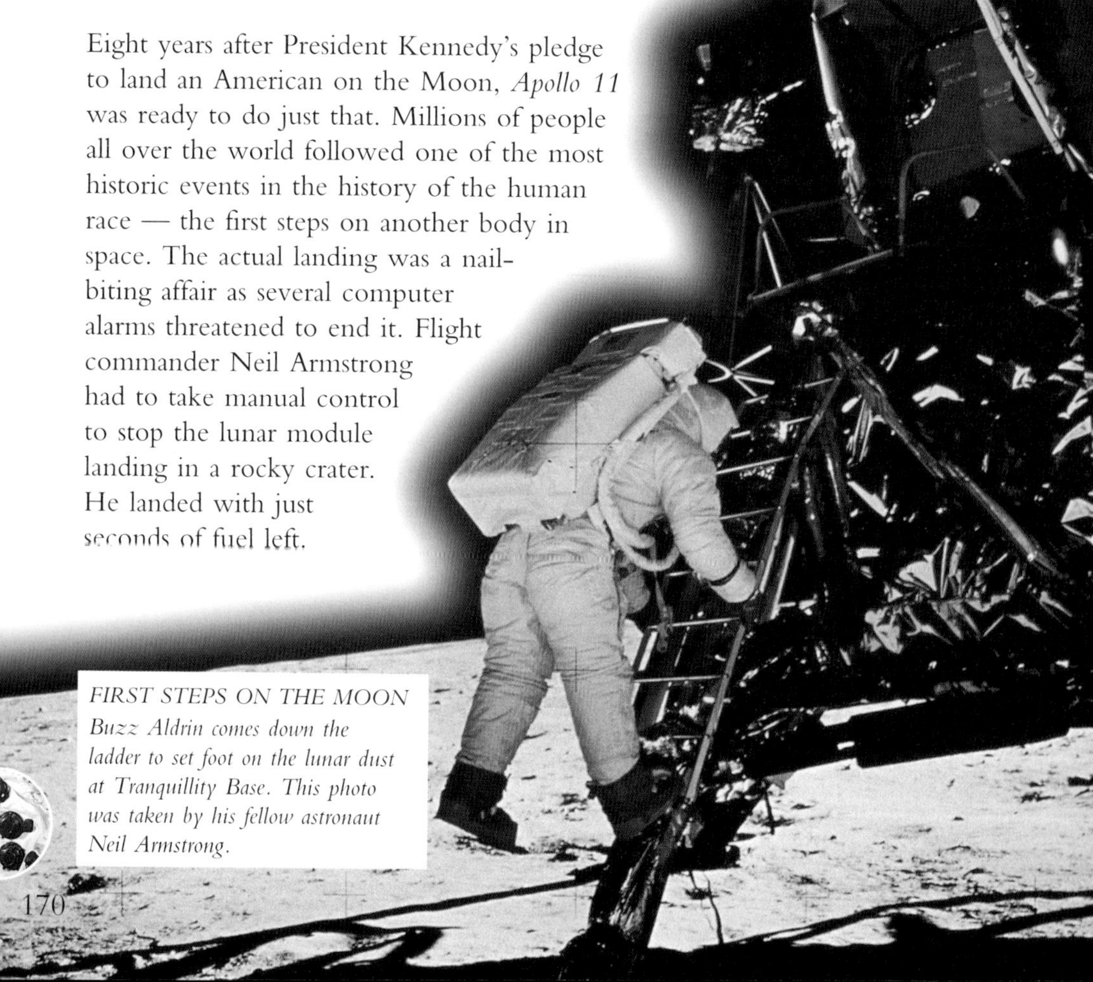

FIRST STEPS ON THE MOON
Buzz Aldrin comes down the ladder to set foot on the lunar dust at Tranquillity Base. This photo was taken by his fellow astronaut Neil Armstrong.

Thanks to satellite technology, millions were able to follow events live on TV. The high point was watching the first walk on the Moon, as Armstrong's ghostly looking figure stepped off the footpad of the lunar module *Eagle*. He was followed by Buzz Aldrin, and both set to work deploying two science instruments and collecting samples of rock to bring back to Earth. The third *Apollo 11* astronaut, Mike Collins, stayed on board the command module.

APOLLO 11'S *CREW*
Neil Armstrong (left), Michael Collins (center) and Buzz Aldrin (right) pose for a formal crew portrait a few weeks before the launch of their epic Apollo 11 *mission to the Moon.*

MOON MISQUOTES

No one knew what Neil Armstrong was going to say when he placed his right boot onto the lunar surface. He only decided finally after he had landed safely. He said that he didn't see the point in worrying about what to say when he didn't know whether he would land successfully — he believed that they had a 50-50 chance of success. Unfortunately, Armstrong's remark became one of the most misquoted in history. He meant to say, "That's one small step for a man, one giant leap for mankind." He actually said, "That's one small step for man, one giant leap for mankind."

SHUTTLES AND STATIONS

When the first Space Shuttle took off in 1981, it marked a major milestone in the history of space exploration. With its ability to land back on Earth like an airplane, the Shuttle is the first manned spacecraft that can be reused. It has made an important contribution to satellite technology. Shuttle astronauts not only deploy communications and other satellites in space, but they even retrieve and repair damaged ones while remaining in space. The Hubble Space

Telescope, which has sent back so many spectacular images of space, was deployed by the Shuttle.

The Space Shuttle has also ferried astronauts to the *Mir* space station, launched by the Soviet Union in 1986. Extra modules have been added to extend the station. *Mir* has received astronauts and scientists from many different countries. The first parts of a new international space station were assembled in space in 1998.

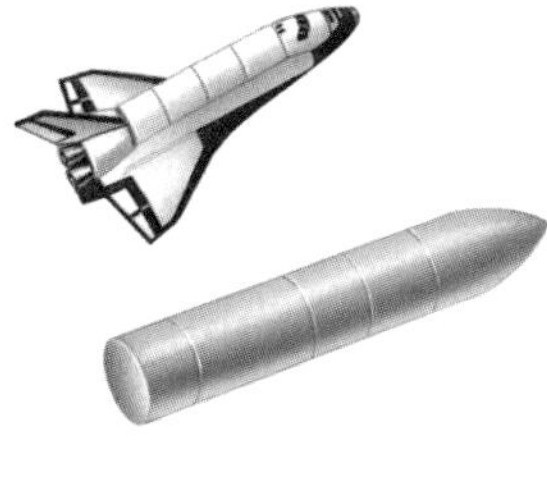

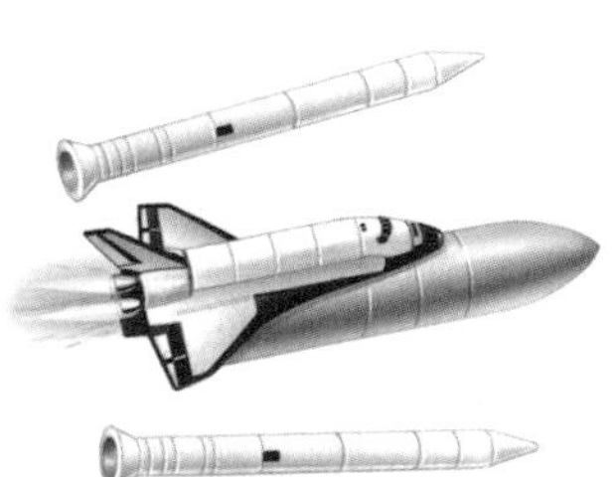

THE SPACE SHUTTLE

The Space Shuttle consists of three main parts: the orbiter spaceplane, two solid rocket boosters, and an external propellant tank. The shuttle is launched using three main engines attached to the orbiter. The engines are fed with propellants from the external tank (ET) and by two solid propellant strap-on solid rocket boosters (SRB). The SRBs use up their fuel after 2 minutes and are ejected. They are recovered for reuse in later flights. The orbiter and its ET continue flying for 6 more minutes until the initial orbit is reached. The ET is then jettisoned. The orbiter's orbital maneuvering system (OMS) engines and reaction control system (RCS) thrusters are used to change the orbit and perform maneuvers.

At the end of the flight, the OMS engines are fired and the orbiter plunges into the Earth's atmosphere at 25 times the speed of sound. The friction causes its 34,000 heat shield tiles to heat up to 2,880 degrees Fahrenheit (1,600 degrees Celsius). The orbiter then lands like a glider.

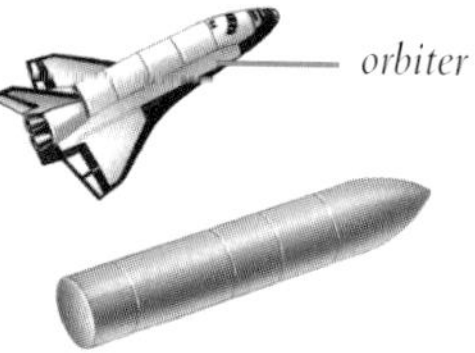

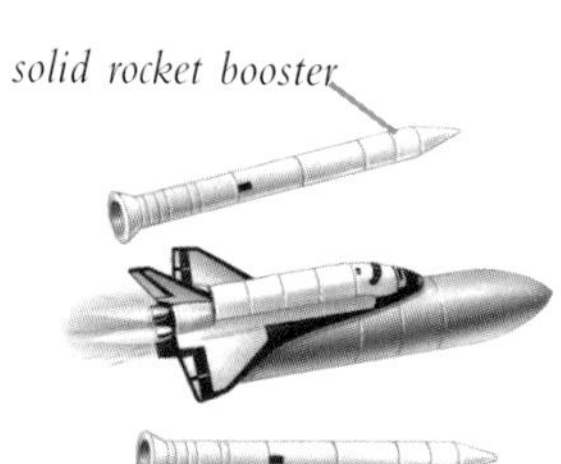

SHUTTLE LAUNCH Two minutes after the Shuttle takes off, the solid rocket boosters are ejected and fall back to Earth by parachute. Six minutes later, the external tank is jettisoned and is destroyed as it reenters the Earth's atmosphere.

ASSEMBLING THE SHUTTLE
The Space Shuttle is assembled vertically inside the Vehicle Assembly Building at the Kennedy Space Center. It is rolled out to the launch pad on a mobile launch platform.

"FLYING BRICK"

The orbiter makes a carefully controlled approach to its landing site. It descends at an angle seven times steeper than that of a normal airliner, with a nose-down angle of 28 degrees. The vehicle finally lands at a speed of more than 200 miles per hour (300 km/h).
The Shuttle has been nicknamed by astronauts as the "flying brick". The commander and pilot of Shuttle missions make hundreds of practice landings using a specially adapted training aircraft.

LANDING OPERATIONS
Immediately after landing, the orbiter is surrounded by a crowd of vehicles and engineers. They prepare it for transfer to the Shuttle Processing Facility where the orbiter is made ready for its next flight.

SKYLAB

Launched in 1973, *Skylab* was America's first and so far only space station — until the launch of the first American section of the International Space Station 25 years later. Much of *Skylab* was made of equipment developed for the Apollo program. Its main orbital workshop comprised a fully equipped empty upper stage of a *Saturn V* rocket. The Apollo telescope mount was based on a leftover lunar module. *Skylab* was launched using a *Saturn V* rocket, and crews were sent to the space station aboard Apollo command service modules that docked with the space station.

Three crews were launched to *Skylab*, consisting of several astronauts who would have flown later Apollo missions that had been canceled. Each crew included one scientist astronaut. The first crew had to repair the space station after it was damaged during launch. The final crew stayed in the space station for 84 days.

SKYLAB *DAMAGE*

Skylab *was damaged during launch and lost one of its solar panels. Another panel was successfully deployed during a brave space walk, saving the whole* Skylab *program.*

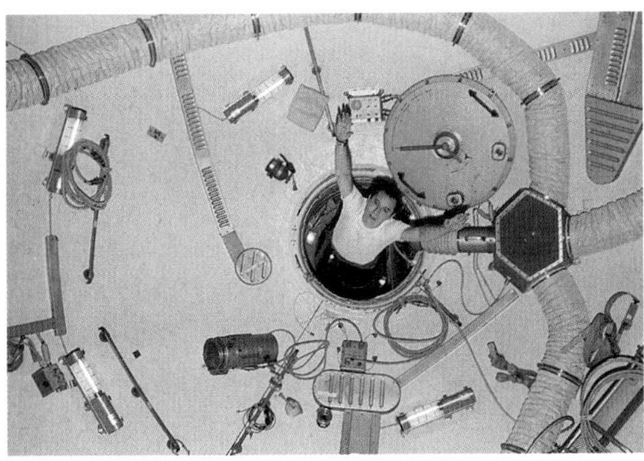

FLOATING AROUND
Scientist astronaut Edward Gibson floats into the orbital workshop during the third and last Skylab *mission in 1973 and 1974. It lasted 84 days, which was an American record.*

DANGEROUS WORK

In May, 1973 former *Apollo 12* astronaut Pete Conrad and scientist Doctor Joe Kerwin performed the bravest and most dangerous space walk in history soon after reaching *Skylab*. They had to prise out the space station's remaining solar array, which was vital to producing power. The other array had been torn off during launch. Conrad and Kerwin used wire cutters to free the jammed panel and pulled it out using pullies that they rigged during the space walk. When it was freed, the panel sprang out and almost hit the space-walking duo.

SCIENTIFIC STUDY
Skylab was a fully equipped science base, with one section devoted to astronomy and the study of the Sun. Here, scientist Gibson mans the operating console for the telescope.

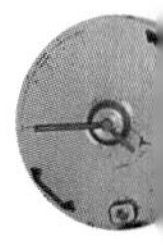

THE MIR SPACE STATION

The *Mir* space station is Russia's extraordinary success story. It was still operating in 1999 13 years after the launch of its first module. The space station consists of the *Mir* core module and a small *Kvant 1* module attached to its rear. Soyuz manned spacecraft and Progress unmanned tankers can dock at a port on the *Kvant 1* module. Attached to the front of the core module is a docking module with five ports. One is used for Soyuz ferries and the other four are for modules that were launched later. The whole space station, with six modules, a Progress tanker and a Soyuz spacecraft, weighs about 130 tons.

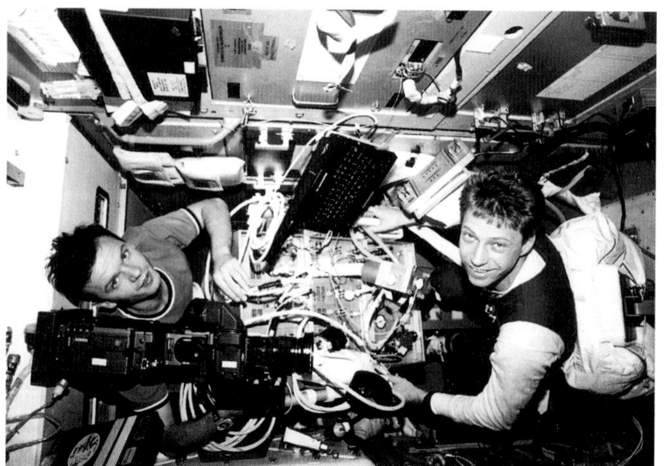

WORKING TOGETHER
Russian cosmonaut Yuri Gidzenko and German astronaut Thomas Reiter at work on board the Mir *space station in 1996. The two men also went on space walks together.*

MIR *OVER NEW ZEALAND*
The first part of the Mir *space station was launched in February 1986, and the final module was added in 1996. The core station was based on earlier Salyut modules.* Mir *is pictured here orbiting over Cape Farewell, on the northern tip of South Island, New Zealand.*

FOREIGN ASTRONAUTS

Thomas Reiter, a German astronaut representing the European Space Agency, flew a 179-day mission on *Mir*. Six American astronauts, including Shannon Lucid and Michael Foale, have also stayed on *Mir* for more than 100 days. France has also flown commercial missions on the space station. One of the French astronauts was Jean Loup Chretien who also flew on *Salyut* 7 — and revisited *Mir* in a Space Shuttle! Shorter missions have been flown by other countries, including Afghanistan and Syria.

Mir has hosted almost 30 main crews, including one cosmonaut who stayed on board for a record 437 days. In addition, many visiting crews have included astronauts from other countries, including the United States, the UK and Germany. Foreign countries have paid Russia for the scientific experiment time on board and for the space flight experience.

Shuttle–Mir Missions

When Russia decided to join the US-led International Space Station, the two countries decided to work together in space — instead of competing as they had done in the past. Arrangements were made for US astronauts to operate on board *Mir,* and for Russian cosmonauts to be launched in the Space Shuttle. Between 1995 and 1998 there were nine joint Shuttle-*Mir* missions. Six US astronauts were delivered by the Shuttle to stay on *Mir* for missions lasting up to 175 days. The first of the astronauts, Norman Thagard, was launched on a Russian Soyuz. Some cosmonauts, including Nikolai Budarin, made their first space flight in the Shuttle.

This was a very productive period of US-Russian cooperation in space. However, it was not without its problems, especially when accidents occurred on *Mir*, including a small fire and a collision.

SHUTTLE MEETS MIR

The space shuttle Atlantis *is seen here alongside* Mir. *This photo was taken by a Russian crew flying in a Soyuz craft around the space station.*

STAYING IN SPACE
Flying his third mission on the Space Shuttle, NASA astronaut Mike Foale was delivered to the Mir *space station for a long stay in space.*

DOCKING COLLISION

Mike Foale was involved in one of the most serious incidents on *Mir*. When an unmanned Progress tanker collided with the station's *Spektr* module, it caused a puncture that threatened to depressurize the station, which would have resulted in the crew's death. Fortunately the leak was not rapid and the crew had time to seal off the module, enabling operations in the rest of the station to continue. Unfortunately most of Foale's possessions and his equipment were in *Spektr* and could not be reached.

FLYING TO MIR
The first Space Shuttle mission to the Mir *space station was launched in June 1995. Its crew included two Russian cosmonauts.*

THE ISS

Since the beginning of the space age, a large space station assembled in space has always been seen as the next logical step in space exploration. Such a space base was not given the go-ahead until 1984. The United States, with its international partners Europe, Canada, and Japan, should have completed the station by 1994. Financial problems have dogged the project for many years, causing the station, called *Freedom*, to be constantly modified. The station was on the point of being canceled in 1994, before anything had been launched. However, the project was saved by the fact that Russia, following the collapse of the Soviet Union, was unable to build its own *Mir 2* station. In 1994, Russia offered

GIANT STATION IN SPACE
When assembly of the International Space Station is completed in 2003, it will be the size of a sports stadium.

BUILT IN SPACE
The first two components of the ISS, the Russian Zarya *module and the American Node 1 module called* Unity*, were joined in orbit in December 1998.*

to join up with NASA to build the International Space Station, or ISS. The project has been continually hit by delays, particularly to Russian components, and the first element of ISS was only launched in 1998.

WORLD'S BIGGEST

The ISS is the biggest international project ever undertaken. When completed, it will have cost more than $30 billion. It is a very ambitious undertaking, particularly because its success will depend on more than 30 Space Shuttle missions over five years and on a series of very complicated in-orbit assembly operations, requiring over 500 hours of space walking. The margin for error is very small. A Space Shuttle failure or another sort of accident would very likely delay the project significantly.

FIRST ISS MISSION
The Space Shuttle Endeavor *blasts off on the first ISS assembly mission on December 4, 1998.* Endeavor *carried the Node attachment module that was linked to the Russian* Zarya *module already in orbit.*

ASTRONAUTS IN SPACE

Astronauts and cosmonauts have been visiting space continuously since Gagarin's historic flight in 1961. Their work during a modern space mission involves a wide range of tasks, from operating spacecrafts and space stations and launching and repairing satellites to carrying out scientific, medical, and engineering experiments in the unusual conditions of space.

An astronaut's life is demanding but also rewarding. After a long

and strenuous training period learning how to cope with the effects of weightlessness, astronauts have to take special care of their bodies while in space. Regular sleep and exercise, and proper eating and drinking, are vital. Today's astronauts no longer have to wear pressurized suits during flight, but once outside their spacecraft they wear spacesuits to protect them from harmful radiation and provide them with oxygen to breathe.

WORKING IN SPACE

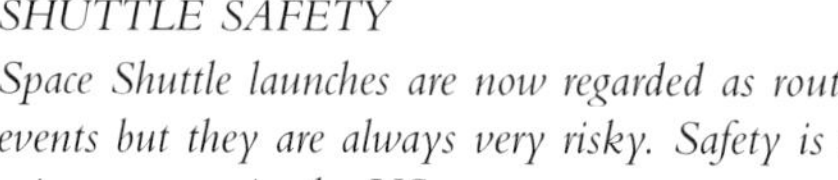

SHUTTLE SAFETY

Space Shuttle launches are now regarded as routine events but they are always very risky. Safety is the prime concern in the US space program.

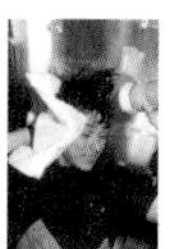

Although every space flight involving humans is risky, traveling into space is now regarded as fairly routine. Astronauts and cosmonauts are not exploring space; they are using it. The days for space exploration are over — space exploitation has begun.

Working is space, however, is totally different. The weightless and airless environment requires a new approach to living and working. It affects everything from eating, drinking and going to the bathroom to working outside the spacecraft in a bulky spacesuit. By the end of 1998, there had been 210 manned space flights, 123 by the United States and 87 by Russia. Each flight has added to our understanding of how to work and live in space more effectively — from the early heroic flights to the largely anonymous missions that take place today.

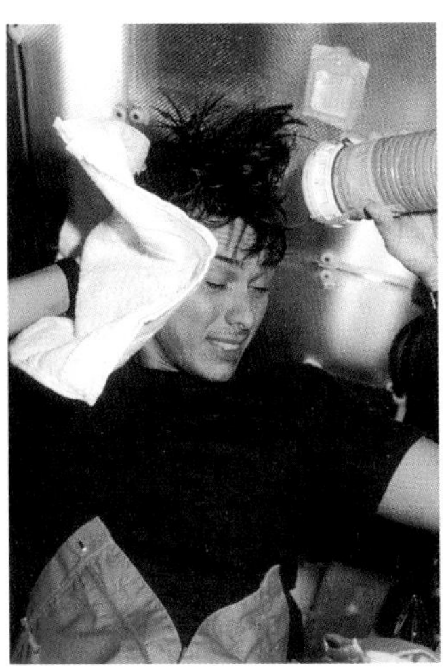

DAILY TASKS
Almost 400 people have now traveled into space since Yuri Gagarin, the first person in space, in 1961. Conditions are now fairly comfortable compared with the early days. Here, astronaut Franklin Chang-Dia, a veteran of six Shuttle missions, is seen drying his hair.

RISKS OF THE JOB

The risk factor in Shuttle launches was illustrated during two launches in 1998. An *Endeavor* launch could have ended in a return to the launch site or something more serious when a pressure sensor in one engine failed. It happened because a broken piece of test equipment had been left in a cooling pipe. A *Discovery* take off caused a scare when the drag chute door became loose and damaged one of the main engines when it fell off during ignition. An on-pad explosion could have occurred if the engine's nozzle had been pierced.

SPACE WALK
Since Russian cosmonaut Alexei Leonov's first walk in space in 1965, there have been more than 140 EVAs, or space walks, in Earth orbit. EVA stands for extravehicular activity.

LIVING IN SPACE

The first space travelers had very basic living conditions. Strapped into relatively small capsules, they had no toilets, eating area, or exercise machine. Food was squeezed into their mouths from toothpaste-like tubes or came in small bite-sized cubes. Astronauts had to pass urine through a tube into a tank. The urine was dumped overboard. Perhaps the most difficult job was passing solid waste into small plastic bags which were then stored for analysis. Space travel was no picnic!

Today, conditions are much better but traveling on the Shuttle is still no more luxurious than a camping trip for seven people in one tent. The Shuttle toilet is similar to the ones on Earth but a vacuum pump has to be used to make sure the waste matter goes down, not up! There is a shower on the *Mir* space station; otherwise astronauts wash using wet wipes. Spacecraft carry exercise machines like treadmills, allowing the crew to take regular time off to keep in shape.

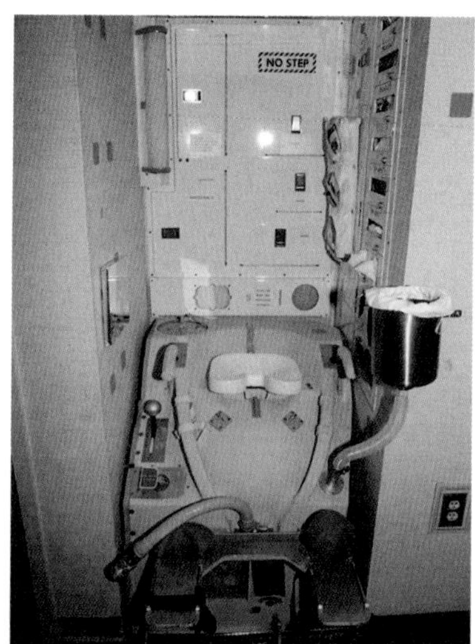

WASTE IN SPACE

Going to the bathroom in space is made more difficult by weightlessness. The first astronauts had no toilets so they had to use plastic bags. The Shuttle has a more conventional-looking toilet (above).

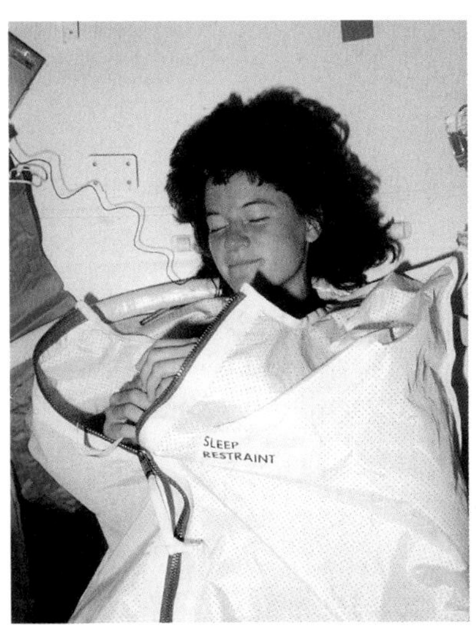

SWEET DREAMS
Most space travelers like to fit themselves snugly into sleeping bags. The bags have to be strapped down to stop them from floating around.

SMELLY SPACECRAFT

Sometimes there have been "housekeeping" accidents in space. Plastic bags for going to the bathroom did not fit properly, or even burst. Some astronauts were sick. Human waste was sometimes left floating around the cabin as it was difficult to clean up. The smell inside the capsule after a long space flight was usually extremely bad! Alan Shepard, the first American in space, faced technical problems before launch. He was delayed for such a long time in his craft that he had to urinate in his pants!

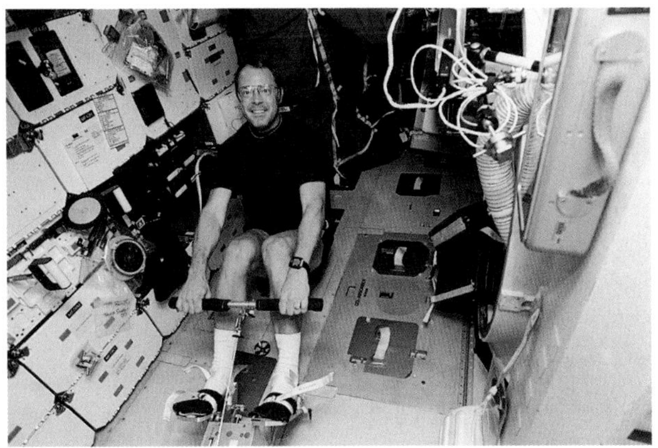

DAILY EXERCISE
To counteract the effects of weightlessness on the human body, each space traveler should exercise for at least 2 hours each day.

SPACESUITS

Astronauts and cosmonauts are launched wearing pressure suits. These protect the crew if the cabin of their spacecraft depressurizes and loses its air supply. The suits are very similar to those worn by fighter pilots. On most missions, the suits are removed once the craft reaches orbit, and put back on when necessary, especially for reentry into the Earth's atmosphere. On early missions, space travelers had to wear the suits throughout the mission — and they were very uncomfortable. Many astronauts got very warm and became dehydrated, losing weight.

A much stronger suit is required for space walking. It needs to protect against the intense heat and cold of space, against radiation from the Sun and possible impacts from tiny meteoroids or space debris dust. These suits have to be strong but very flexible, allowing astronauts to work effectively.

OUTSIDE THE SHUTTLE
Astronaut Kathryn Thornton is seen outside the Space Shuttle repairing the Hubble Space Telescope. She is wearing the Environmental Mobility Unit spacesuit with an integrated portable life-support system backpack.

SUITS FOR SAFETY
Like high-altitude aircraft pilots, the Shuttle astronauts wear ascent and descent pressure suits like this one, in case the cabin is depressurized.

SUITS OR NOT

It seems amazing that some astronauts and cosmonauts have flown into space without pressure suits. Some Russian Voskhod and Soyuz pilots wore tracksuits, while Shuttle astronauts from 1982 to 1986 wore flight overalls and helmets with an emergency oxygen supply to provide basic protection against depressurization. The last Shuttle astronauts to wear overalls were the seven crew members on the fatal flight of the *Challenger* in 1986.

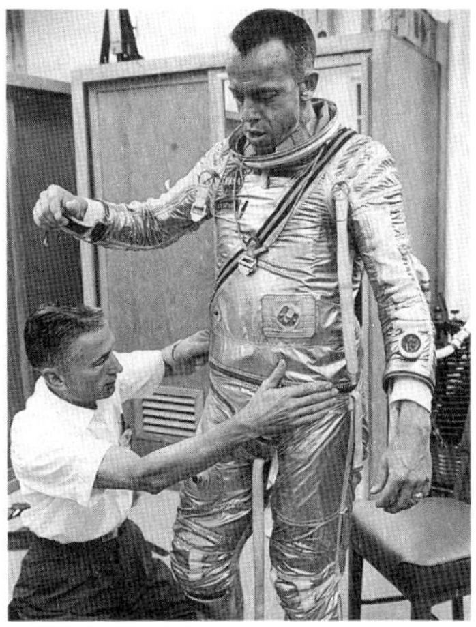

EARLY SPACESUITS
Early astronauts had to wear their spacesuits throughout the whole mission. The suits had to be comfortable, flexible enough to allow movement and at the same time provide protection from the hazards of space.

WEIGHTLESSNESS

Weightlessness — or zero gas it is sometimes called — is strictly speaking not a correct term. There is still a very small amount of gravity at work in orbit. The term most commonly used today is microgravity. For space travelers, this is an alien environment. Its immediate effects are that blood tends to "float," pooling more in the face and causing puffiness. Mucus in the head does not drain and the astronaut becomes a bit congested. Often a space traveler feels sick for a day or two at first because the delicate parts of the inner ear that control balance are affected. It's a bit like being seasick. Longer-term

FLOATING IS FUN
German scientist Ulrich Walter floats inside a Spacelab *module on board the Space Shuttle. Working in space can be fun.*

effects are a loss of calcium in the bones and muscle wastage — this even affects the heart. These side-effects would be a disaster if they were not overcome with the help of medicines and exercise. Space travelers are urged to spend at least two hours a day doing strenuous exercise.

FREE FALL

When a car goes quickly over a humpback bridge, its occupants experience very briefly what it is like to be weightless. A parachutist experiences weightlessness before the parachute opens after jumping from an aircraft. Likewise, astronauts in orbit are in free fall so there is no opposing force of gravity to give the feeling of weight. But there is still the small mutual gravitational attraction that exists between objects, for example between an astronaut and a bag of food or a pen. So the term "microgravity" is used to describe this.

ASTRONAUT RETURNS TO SPACE
John Glenn — the first American to orbit the Earth in 1962 — is shown here on board the Space Shuttle in 1998, aged 77. He underwent intensive medical checks during the flight.

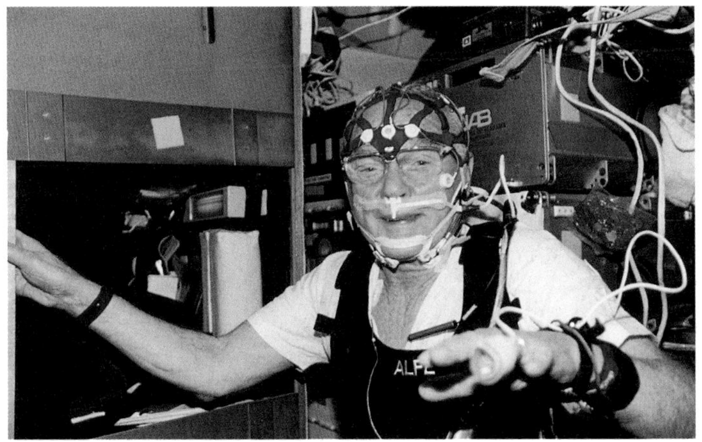

ALL IN A DAY'S WORK

An astronaut flying an intensive 10-day Space Shuttle mission requires a different work approach to one completing a 100-day shift on a space station. While cosmonauts on the *Mir* space station have their own daily tasks, such as conducting experiments and repairing equipment, they can decide for themselves how to complete the task and how long to take. A Shuttle mission packs in as much as possible into the short time in orbit. Each crew member has a checklist that indicates, minute by minute, what jobs have to be done, when meals can be taken and when the crew can sleep. Problems do occur and so the checklists have to be amended.

Although time is set aside for rest and exercise, the work is very intensive, with up to seven people working together in a confined space. They may have trained together for months but there are still occasional moments of disagreement and stress.

FLYING THE SHUTTLE
The most intense parts of a Shuttle mission are the launch and landing. The pilot and commander have to be prepared at all times for emergencies.

WORKING THE ROBOT ARM
The Shuttle's robot arm is delicately moved using a hand controller and computers inside the Shuttle's cockpit.

HELP FROM EARTH

A Shuttle crew is supported by a ground-based team of hundreds of people. Mission Control works directly with the crew via a capsule communicator, or capcom, who is a fellow astronaut. Communications with Mission Control are made directly or via a fleet of tracking and data relay satellites, wherever the Shuttle may be in orbit around the Earth. Engineers in charge of equipment and experiments are always on hand to offer advice. Often the same experiment that is being operated in space is carried out simultaneously on the ground in order to compare the results.

STORAGE SPACE
Hundreds of lockers hold equipment and clothes for the crew of the Space Shuttle. The addition of the Spacehab module shown here has enabled more equipment to be carried.

SPACE WALKING

All space missions today carry special spacesuits for EVAs. Even if space walks are not planned, sometimes they are necessary in emergencies, for example to repair a broken piece of equipment. EVAs have become fairly routine events today. Since Alexei Leonov floated outside *Voskhod 2* in March 1965 on the first EVA, there have been almost 200 in Earth orbit, on the Moon, and in between the Moon and the Earth.

The record for the longest EVA — and the first and so far only EVA made by three astronauts together — is over 8 hours in May 1992. These astronauts were making an emergency EVA to retrieve a stranded satellite. The longest Moon walk, by Gene Cernan and Jack Schmitt of *Apollo 17* in 1972, lasted 7 hours, 37 minutes. Three astronauts have made trans-Earth EVAs during Apollo missions on the way home from the Moon. The Hubble Space Telescope has been repaired and serviced on two Shuttle missions involving about 60 hours of space walking. Another servicing mission is planned for 2000.

WEARING A MMU

The first independent space walks without using tethers were made by Bruce McCandless and Robert Stewart in 1984 wearing manned maneuvring unit (MMU) backpacks.

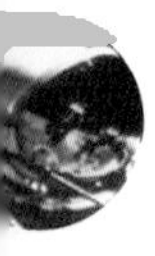

FIRST AMERICAN TO WALK IN SPACE

The first American astronaut to walk in space was Edward White of Gemini 4 *in June 1965. White remained outside the spacecraft for 22 minutes.*

WHAT'S AN MMU?

The Manned Maneuvring Unit (MMU) is a self-contained backpack that can be attached to the Shuttle spacesuit. Nitrogen gas is propelled from 24 nozzles positioned around its exterior to move the astronaut up and down, from side to side and forward and backward at a touch of the hand controller. The MMU was used nine times on three Space Shuttle missions in 1984. The longest excursion, during the first test, lasted 1 hour, 22 minutes.

MOVING AROUND IN SPACE

A Shuttle astronaut on a space walk is maneuvered around the payload bay. The astronaut is standing in footholds at the end of the Shuttle's remote manipulator system robot arm.

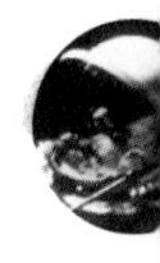

FAMOUS MISSIONS 1

After Yuri Gagarin's first historic mission, many firsts in space were notched up in the early days of space flight. They included Gherman Titov's first sleep in space during the first daylong flight in August 1961. Valentina Tereshkova became the first female in space aboard *Vostok 6* in 1963. Another milestone was the first space docking between *Gemini 8*, commanded by Neil Armstrong, and an Agena target rocket in March 1966. Armstrong, of course, was the first man on the Moon three years later.

The first transfer of crews between two spacecrafts, the Russian *Soyuz 5* and *4*, was made in January 1969. Veteran Gemini, Apollo, and Shuttle astronaut John Young became the first

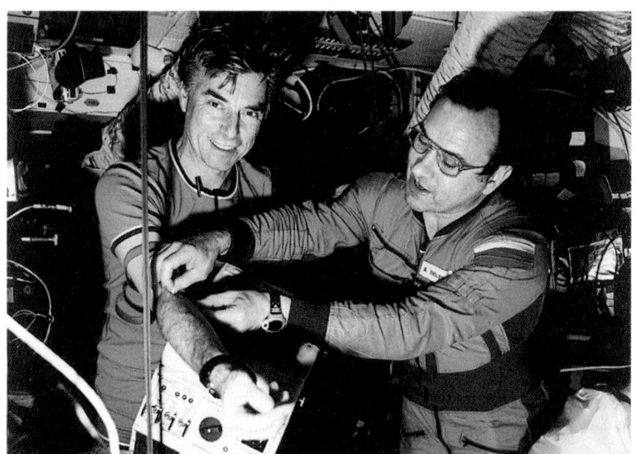

RECORD SPACE FLIGHT
The longest space flight of 437 days was made by Russian doctor Valeri Poliakov (right), between January 1994 and March 1995. Poliakov also made a 240-day flight in 1988–1989, bringing his total time spent in space to a record 678 days, 16 hours.

person to make six space flights on a November 1983 mission of the Shuttle. The biggest crew on board at launch was eight members on a Shuttle mission in October 1985. The first crew to be launched in its entirety a second time were the seven astronauts of Shuttle flight STS 83. It was aborted after 3 days and relaunched 88 days later.

SPACE ENDURANCE

There are unlikely to be any space flights longer than Poliakov's 437 days for a number of years, because shifts aboard the International Space Station are likely to last about 90 days. The longest US space flight, of 188 days, was made by Shannon Lucid in 1995–1996, during which she spent most of her time on the Russian *Mir* space station. Lucid's flight is the 13th longest. There have been three Russian flights over 300 days, four more than 200 days and four more than 100 days. The first space flight to last over a year was Soyuz TM4 between 1987 and 1988.

GEMINI RENDEZVOUS

Gemini 6, *pictured here, performed the first rendezvous in space with a target,* Gemini 7, *on December 15,1965. It was a major milestone in space exploration.*

FAMOUS MISSIONS 2

The first Space Shuttle mission, made by veteran astronaut John Young and pilot Bob Crippen in April 1981, was a historic test flight. It involved a totally untried and untested space plane and launch system. Its success made possible many famous missions, including the one in which the Solar Max satellite was captured, repaired, and redeployed in working order.

In 1984 Shuttle astronauts captured and brought back to Earth two communications satellites, to be repaired and relaunched.

A similar mission was performed in 1985 when a communications satellite was captured, repaired and redeployed. Shuttle mission STS 49 in 1992 involved an extraordinary feat in which three astronauts captured a stranded Intelsat communications satellite the size of a house, fitted it with a new rocket motor, and sent it back into orbit. Later, the engine fired and the satellite entered service in geostationary orbit.

HERO RETURNS TO SPACE
STS 95 would have been a rather routine Space Shuttle mission had it not included 77-year-old space hero and grandfather John Glenn (second from right)among its crew.

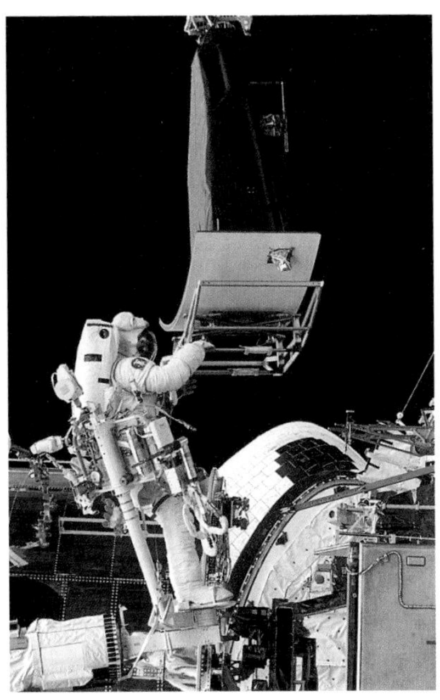

HUBBLE REPAIR
Jeff Hoffman works on the wide-field and planetary camera of the Hubble Space Telescope during the first Shuttle mission to repair and service the telescope.

GUESTS IN SPACE

Many Space Shuttle missions have carried non-astronaut payload specialists. At first these were qualified scientists, but by 1985 passengers such as government officials were being carried. NASA was criticized for this, especially after the *Challenger* disaster in 1986, which killed seven crew members including a teacher. A special case was made when 77-year-old John Glenn, the first American in orbit and a well-respected US senator, was given a seat on a Shuttle mission in 1998.

CAPTURING SOLAR MAX
Astronaut James van Hoften tests his MMU during the 1984 Shuttle mission in which the Solar Max satellite was retrieved, repaired, and redeployed.

THE FUTURE IN SPACE

The US Space Shuttle was to have flown 50 times a year, making space travel routine in the 1980s. In 1970, NASA was still planning a manned flight to Mars for the 1980s. We read about different proposals for new vehicles and projects to make space travel routine, and to make manned flights to Mars a reality, but these should be treated with caution. The two main problems are technology and money. The technology has been much more difficult to prove, and there have not been enough funds. A manned flight to Mars is unlikely before 2015.

US SPACE PLANE
This plane was once the American dream — the National Aerospace Space Plane which would make traveling in space as routine as today's air travel.

A routine space plane taking tourists into space could fly within 5 years but would be incredibly expensive, with the cost of a ticket running into millions of dollars. Airliner-type space travel is unlikely for at least another 20 years.

LEGACY OF APOLLO

There is one factor that may change the pace of today's space program – the Apollo legacy. The race to the Moon resulted from a unique combination of political will, a quest for exploration, and the available technology. It could happen again. For example, if Martian soil samples do prove to have some form of life, a manned Mars program would be given the go-ahead. If the Earth's environment degrades so much that we have to look for another place to live, the focus of space exploration would change dramatically.

VENTURESTAR
The VentureStar project shown here is more likely to happen in the near future. However, it still depends on billions of dollars of private investment and on new technology that has still to be proved.

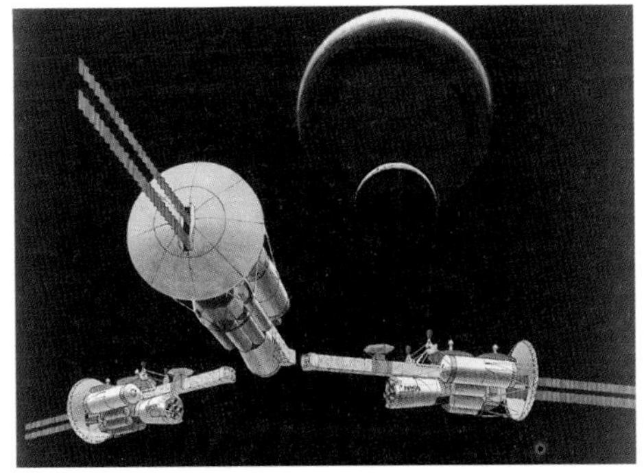

IS THERE LIFE ON MARS?
This spacecraft could return samples from the surface of Mars to Earth for analysis in 2008. The question "Is there life on Mars?" might be answered finally.

STAR CHARTS

On a clear night, when the sky is dark and there is no Full Moon, you should be able to pick out some of the more obvious constellations and star patterns in the sky.

The constellations that you can see depend on where you are in the world, and what time of year it is. On the following pages in this section you will find a series of star charts depicting the more obvious stars and star patterns that you might find. The charts are divided into two parts: those for the northern hemisphere

and those for the southern hemisphere. For each of the four seasons (spring, summer, autumn, winter) there is a pair of charts — the one you use depends on whether you are facing to the north or to the south.

The white dots on the charts represent individual stars. The bigger the dot, the brighter the star is. The lines that link the stars in a particular constellation have been added to help you identify the stars and star groups. You will also find the Milky Way indicated on each chart.

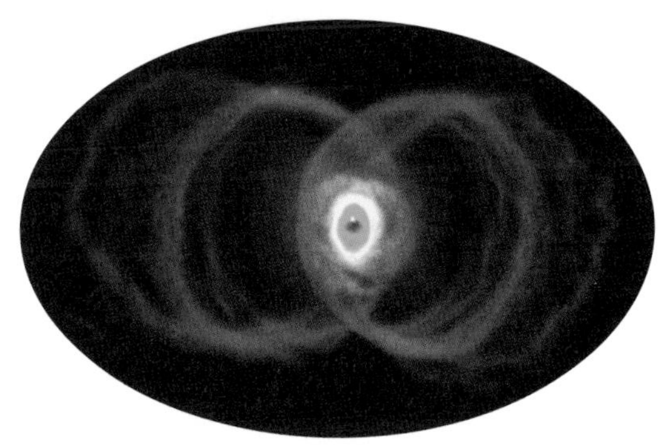

SPRING: LOOKING NORTH

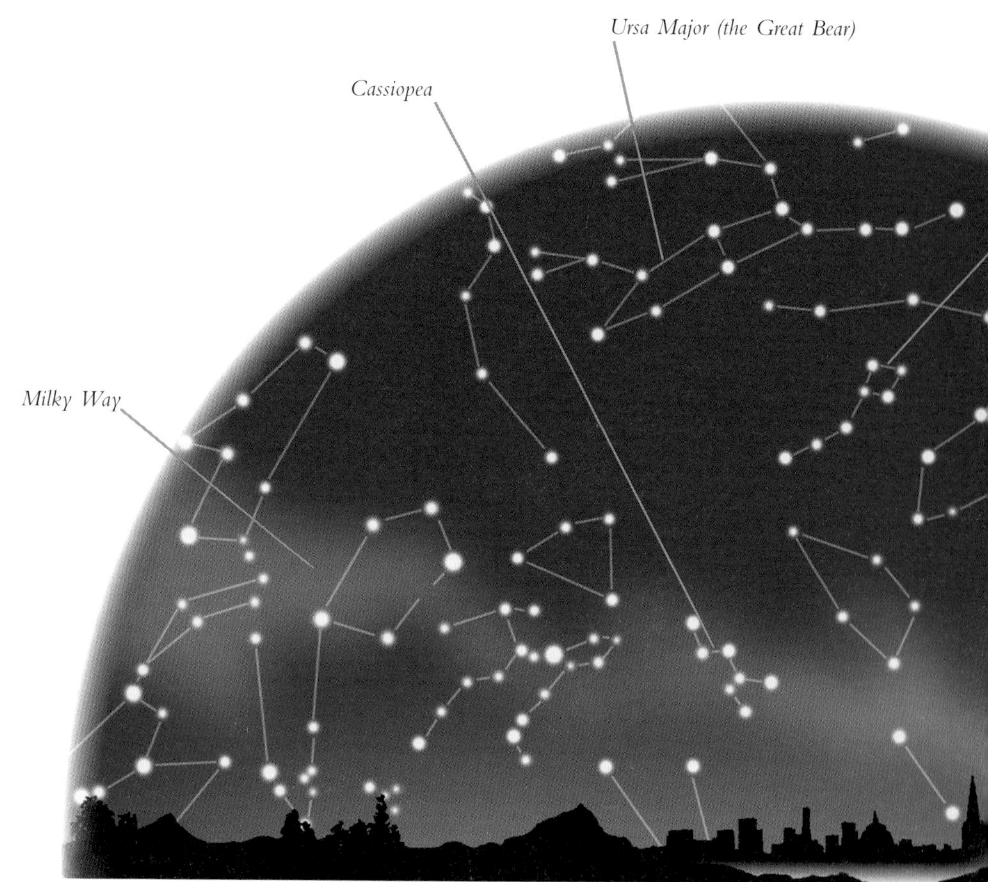

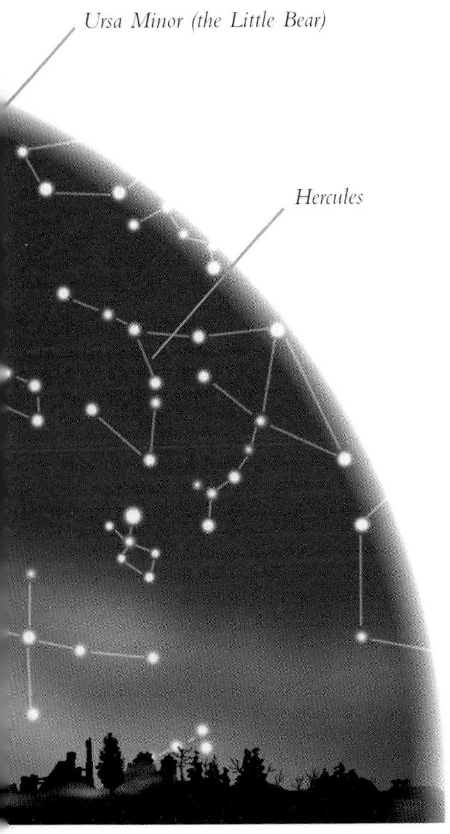

APRIL

The famous "W" shape of Cassiopea can be seen easily without having to crane your neck! The constellation is flooded by the Milky Way.

Hercules is a sprawling constellation dominated by a square shape. If you follow the the line of stars beneath the bottom left star, you will see a "red" star called Ras Algethi. This is a huge red supergiant star, which could be the largest known.

Spring: Looking South

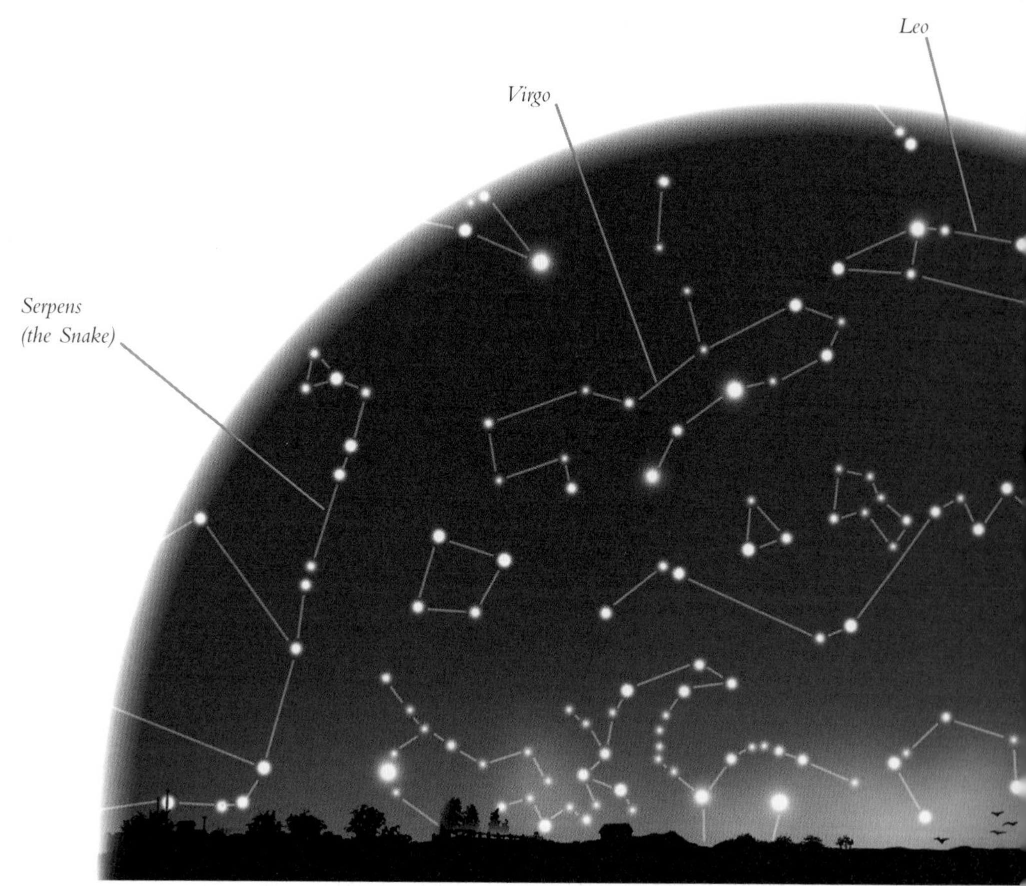

APRIL

The large constellation Virgo is dominated by a bright, blue-white star called Spica, one of the brightest in the night sky. This is 220 light-years away from Earth.

The constellation Hydra (the Water Monster) is the largest constellation, although it is not all that easy to make out. It forms a long snakelike path across the sky.

Another major constellation is Leo (the Lion). To the right of the inverted "question mark", you can see the Praesepe star cluster.

SUMMER: LOOKING NORTH

JULY

The summer is a good time to see the Big Dipper at its best. Ursa Major (the Great Bear) is a great pointer in the sky.

The easiest way to find Polaris, the Pole Star, is to run an imaginary straight line upward from the two stars on the right side of the 'basin" shape part of the Big Dipper. The "bent handle" on the left of the Big Dipper points directly to Arcturus, the reddish-orange star in the constellation Boötes. Arcturus is 40 light-years away.

SUMMER: LOOKING SOUTH

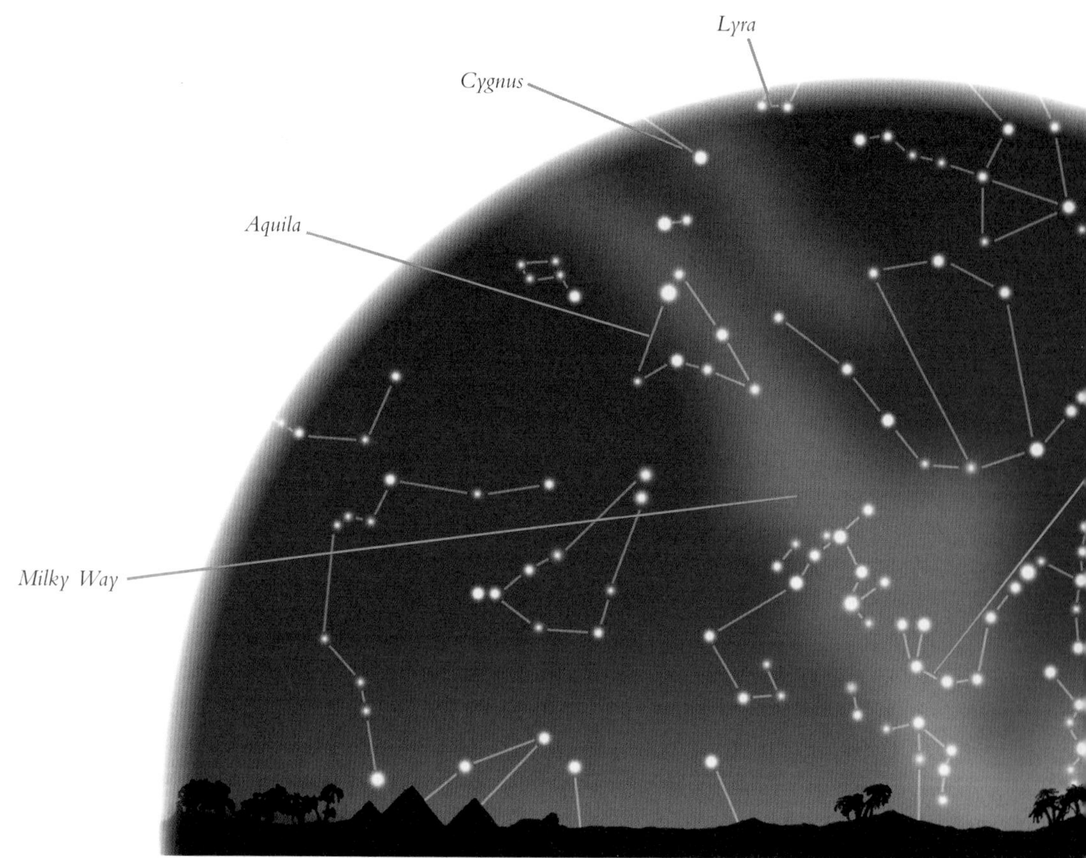

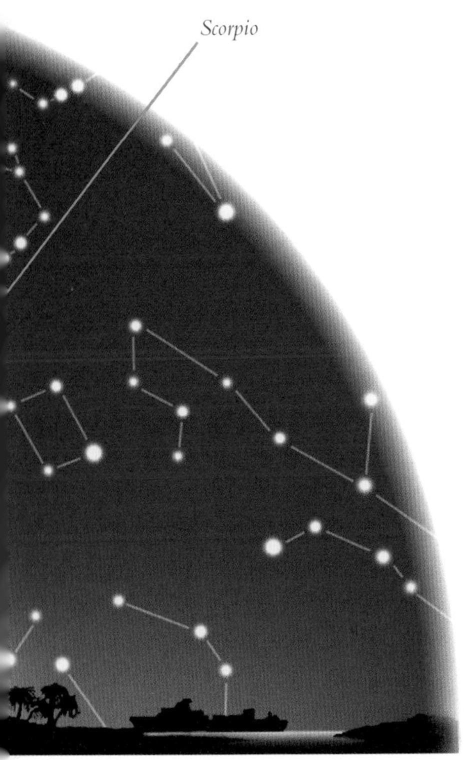

JULY

The summer sky is dominated by a triangle of bright stars: Deneb, Altair, and Vega. Deneb is the main star of what is sometimes called the northern cross — it is actually the constellation Cygnus (the Swan). Altair is part of a constellation called Aquila (the Eagle), while Vega, one of the brightest stars in the sky, is in the constellation Lyra (the Harp).

Another easily recognizable constellation at this time of year is Scorpio (the Scorpion). At the "head" of this scorpion-shaped constellation is a red giant star called Antares.

AUTUMN: LOOKING NORTH

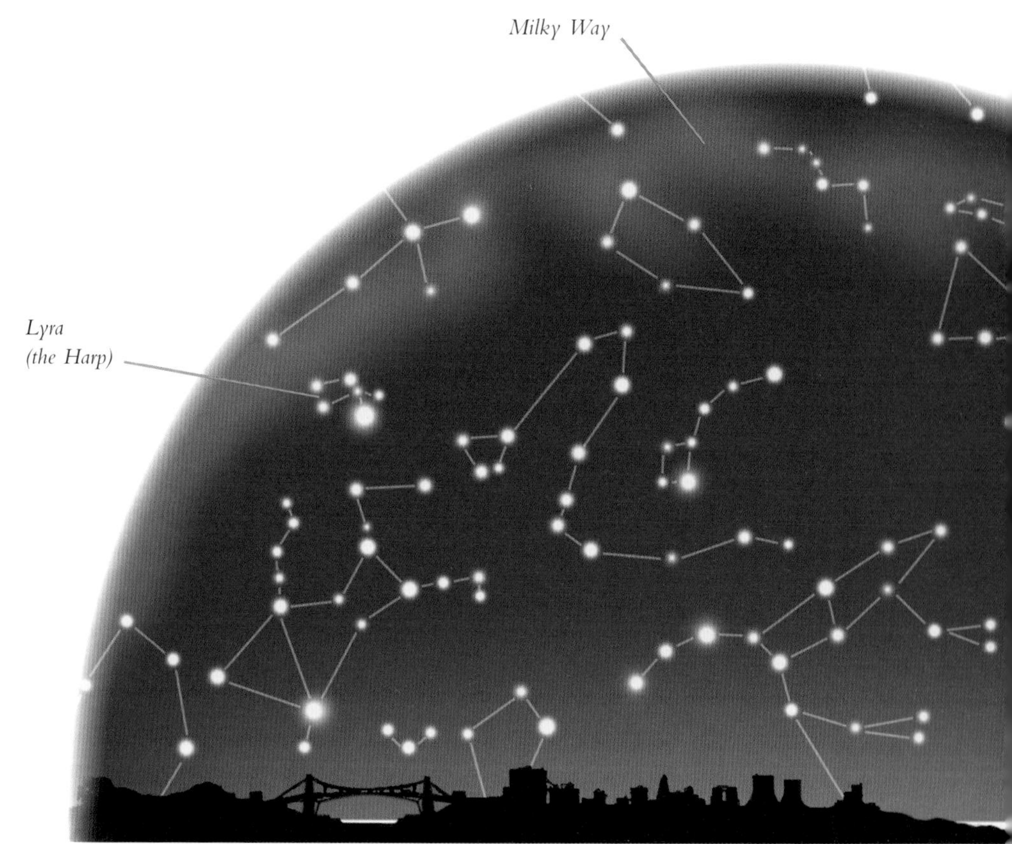

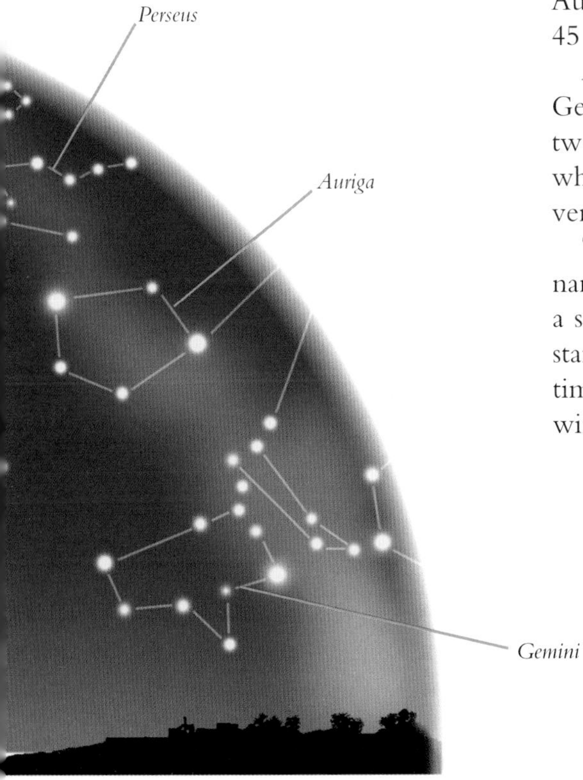

Perseus

Auriga

Gemini

OCTOBER

The famous winter stars start to make their appearance in October. One of the first to appear is Capella, the main and bright yellow star of the constellation Auriga (the Charioteer). Capella is 45 light-years away.

Another famous constellation is Gemini (the Twins). It is based on two bright stars, Castor and Pollux, which used to be the same brightness very many years ago.

The constellation Perseus, which is named for a mythological figure, has a star called Algol. It is called a variable star because its brightness changes over time. Algol is sometimes known as the winking star.

AUTUMN: LOOKING SOUTH

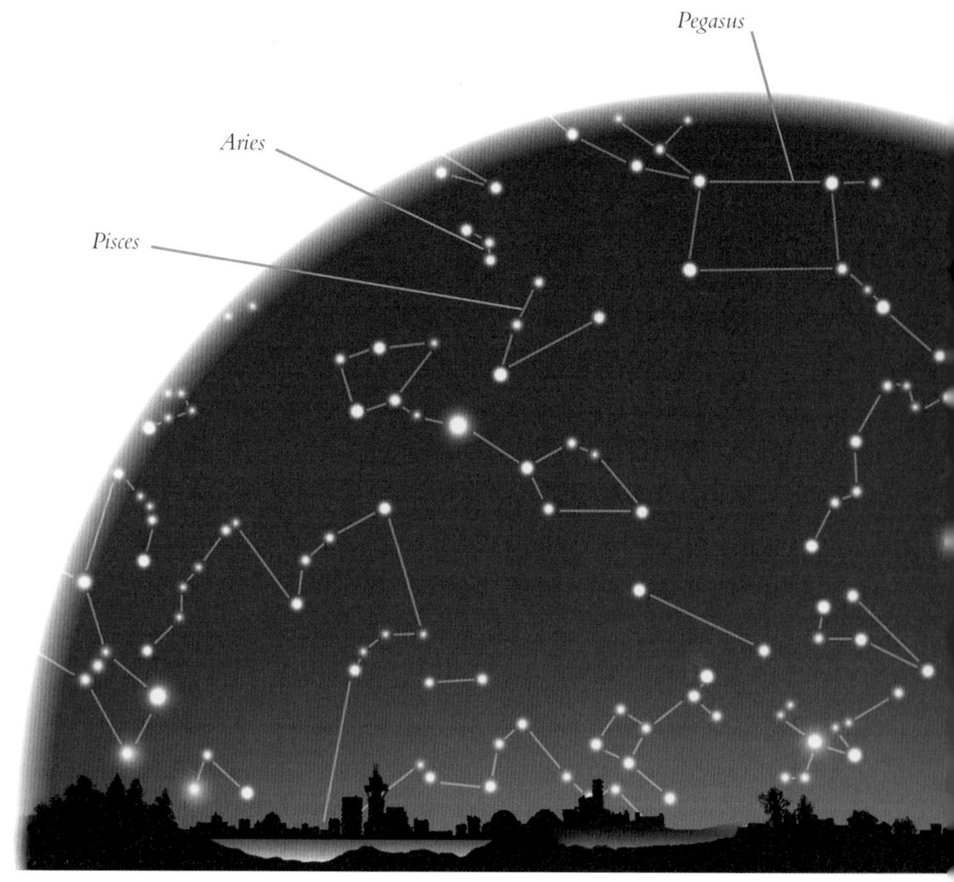

OCTOBER

The square of Pegasus (the Winged Horse) is seen easily in October. Beneath it, to the right, are the constellations Capricornus (the Sea Goat) and Aquarius (the Water Bearer). Both are constellations of the Zodiac, the group of star patterns which lie in the path along which the Sun passes.

If you follow a curve upward to the left you will see the constellations Pisces (the Fishes) and Aries (the Ram), which are also part of the Zodiac.

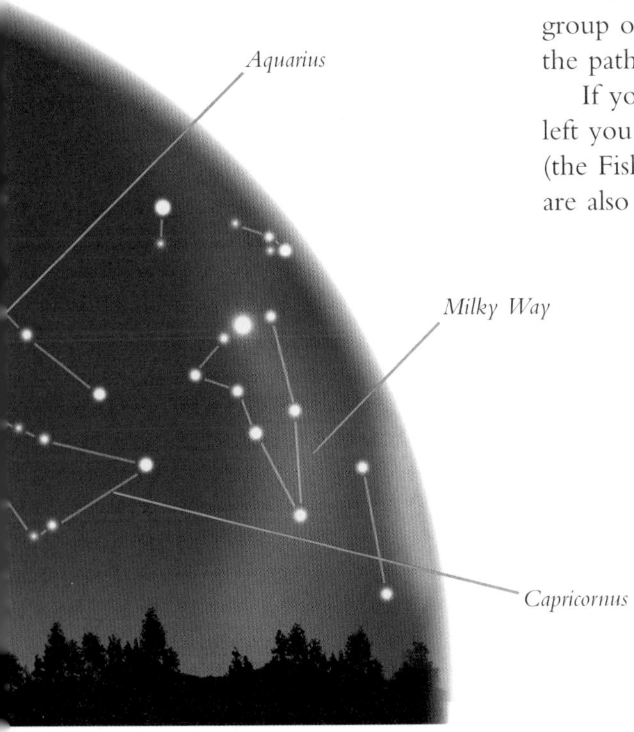

WINTER: LOOKING NORTH

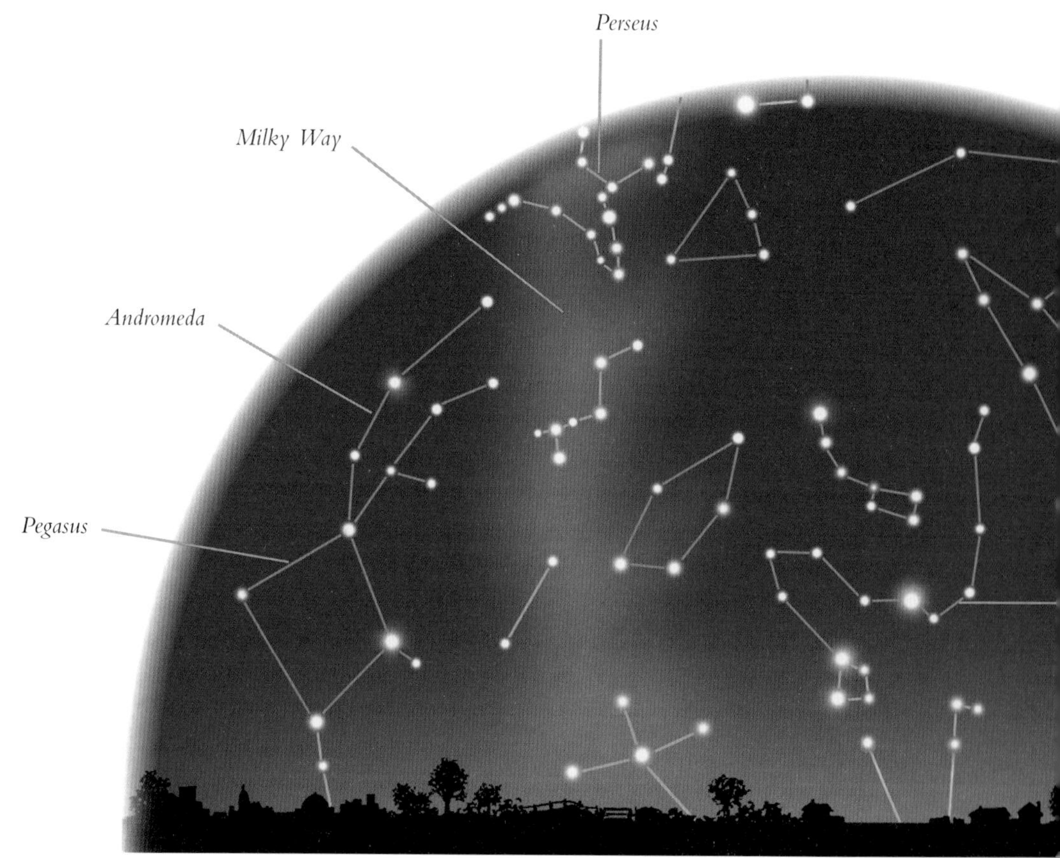

JANUARY

Winter could be the best time to see the famous Andromeda Galaxy. The sky is at its darkest, and on the cold winter nights it can be crystal clear. Following a line of stars up from the top left of the square of Pegasus (the Winged Horse) is a line of stars that makes up the constellation Andromeda. The second of the three brighter stars in this line is a "sighter"'star to help you see the Andromeda Galaxy, upward to the right. The galaxy can be seen with the naked eye.

WINTER: LOOKING SOUTH

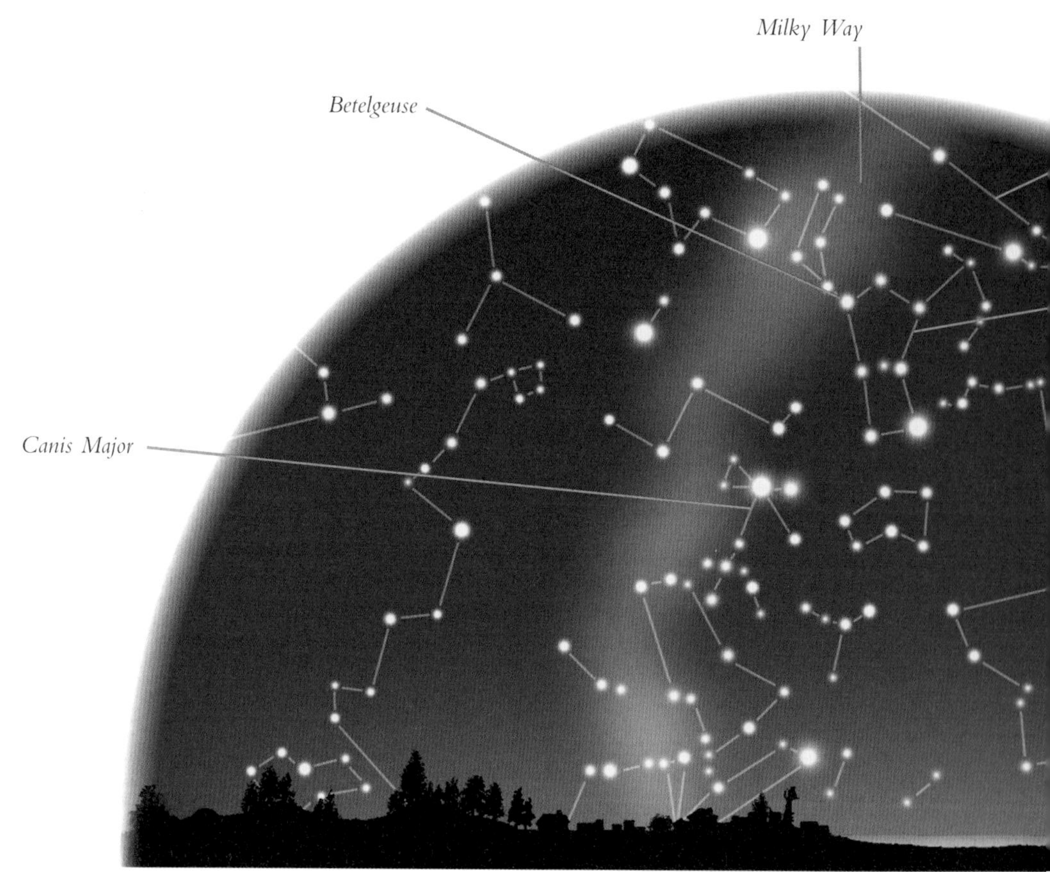

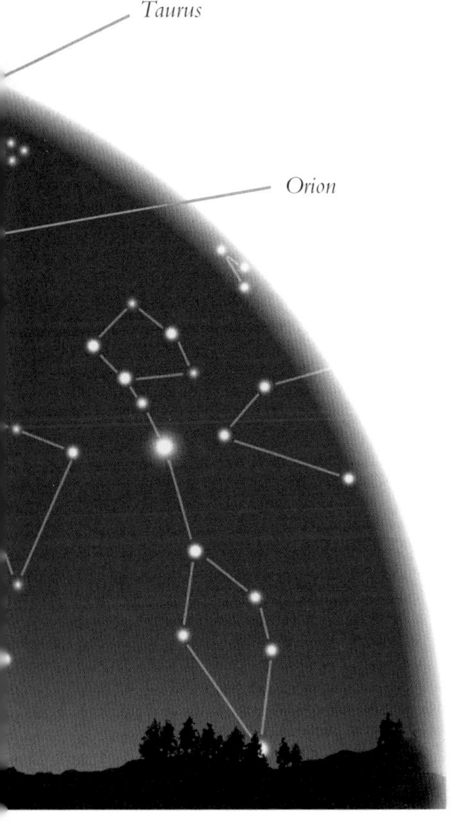

JANUARY

The winter stars are in their splendor at this time of the year, particularly Orion and Taurus. Other important sights in the January skies are the constellation Canis Major (the Big Dog) and Sirius, the brightest star in the heavens.

Orion (the Hunter) is dominated by the "belt" of three stars with its "sword". In the sword is the famous Orion Nebula. The bright red star at the top left of Orion is Betelgeuse, and the bright white star at the bottom right is Rigel. On the right you can make out the hunter's "bow," which is aiming an imaginary arrow at Taurus (the Bull).

SPRING: LOOKING NORTH

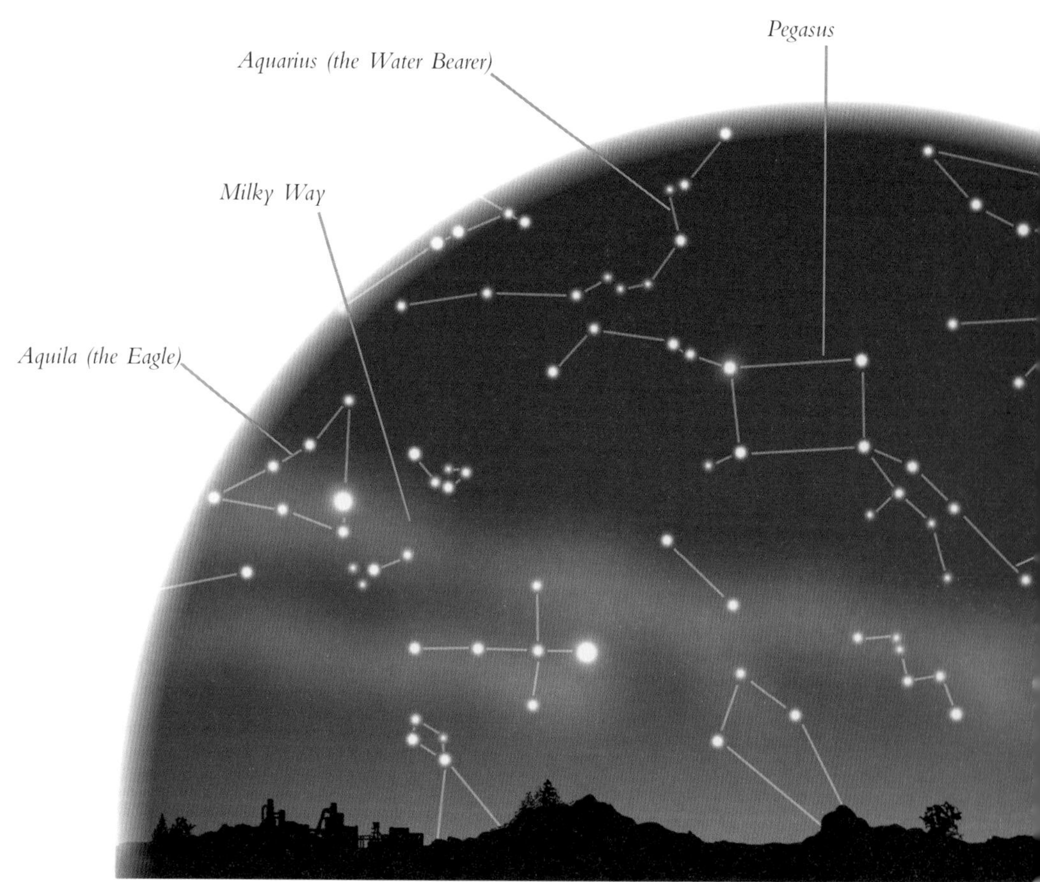

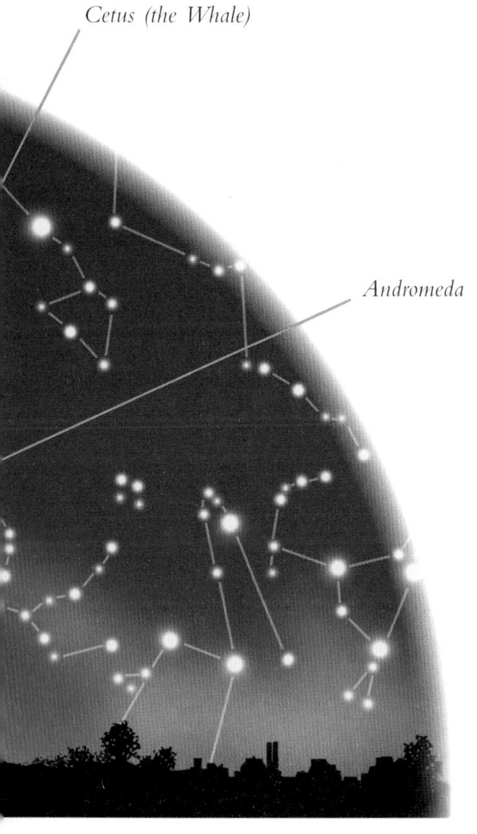

OCTOBER

The spring sky is dominated by the enormous square of Pegasus (the Winged Horse). (Pegasus was an immortal winged horse in Greek mythology.)

In the top left corner of the square of Pegasus is the star Markab. If you follow a line of stars up from Markab you will see another line of stars that make up the constellation Andromeda. The second of three brighter stars in this line is a "sighter" to see the famous Andromeda Galaxy, which is upward to the right.

SPRING: LOOKING SOUTH

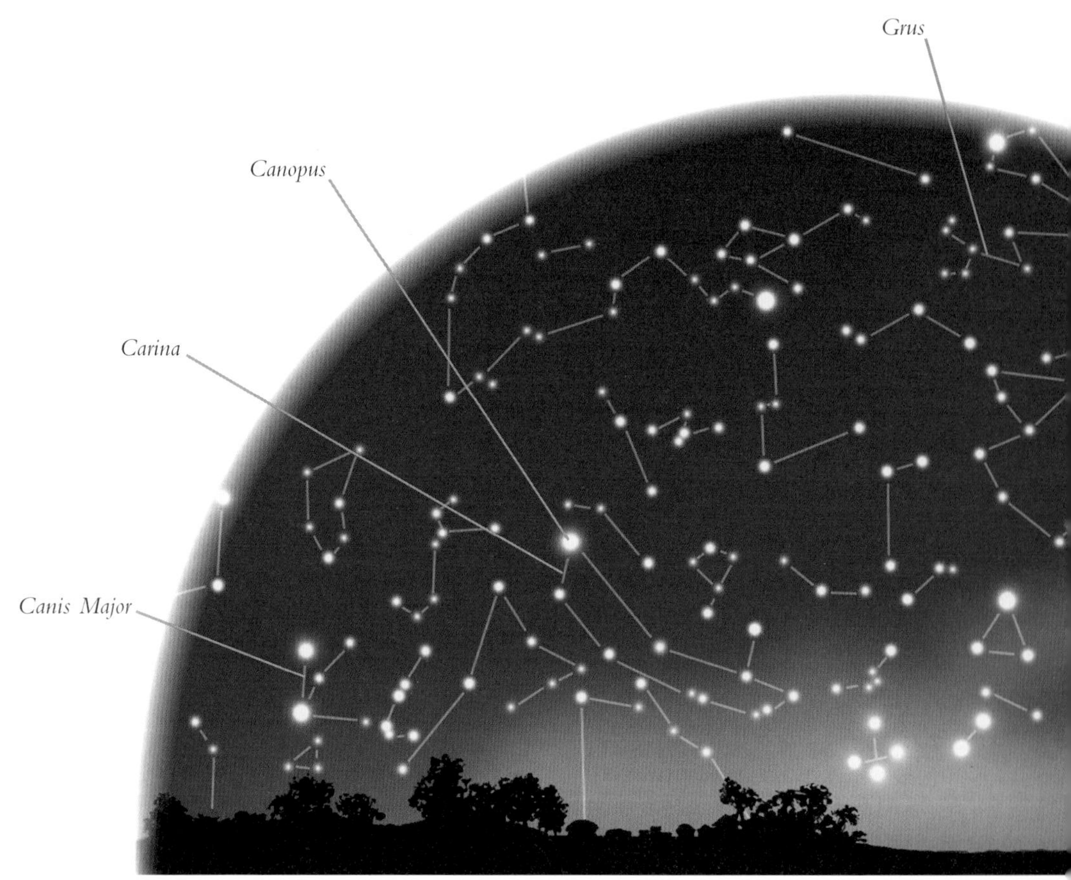

OCTOBER

The two stars that are very close together in the constellation Grus (the Crane) are a double-star system. They can be seen with the naked eye.

To the left of the sky is the brightest star, Sirius, in the constellation Canis Major (the Big Dog). The second brightest star, Canopus, is just above to the right, in the constellation Carina (the Ship's Keel). Canopus may be 80,000 times more luminous than the Sun.

Near the curling "sting" in the tail of Scorpio (the Scorpion) is a superb cluster of stars which can be seen with the naked eye.

SUMMER: LOOKING NORTH

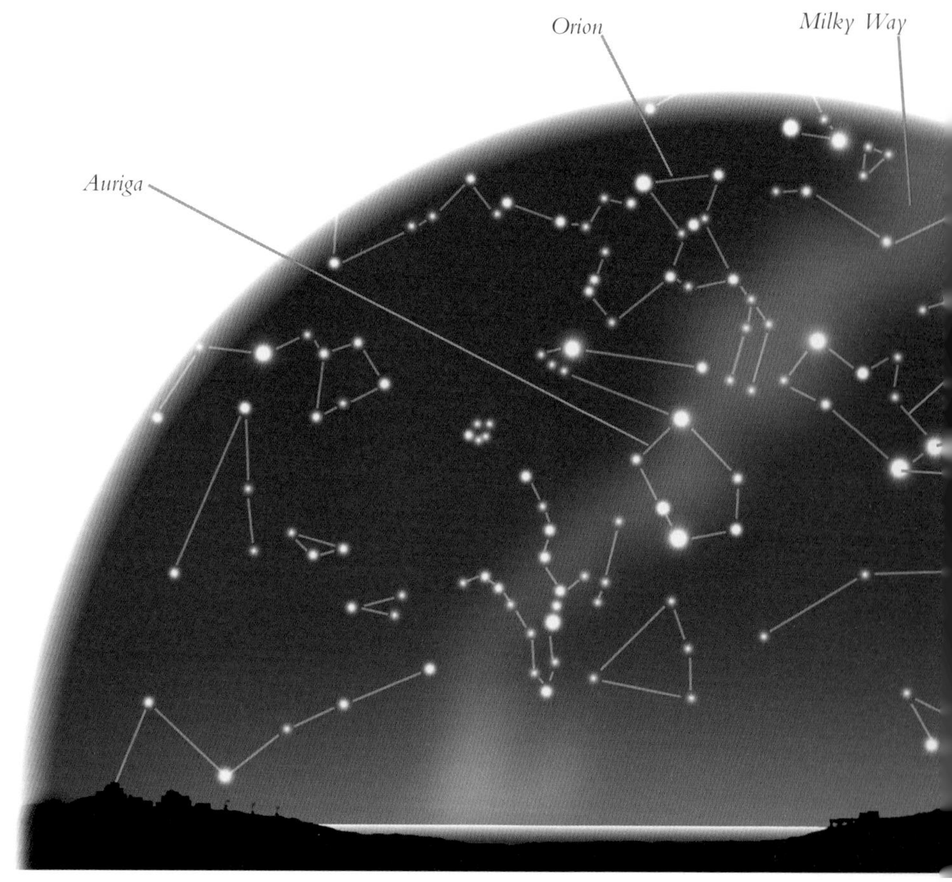

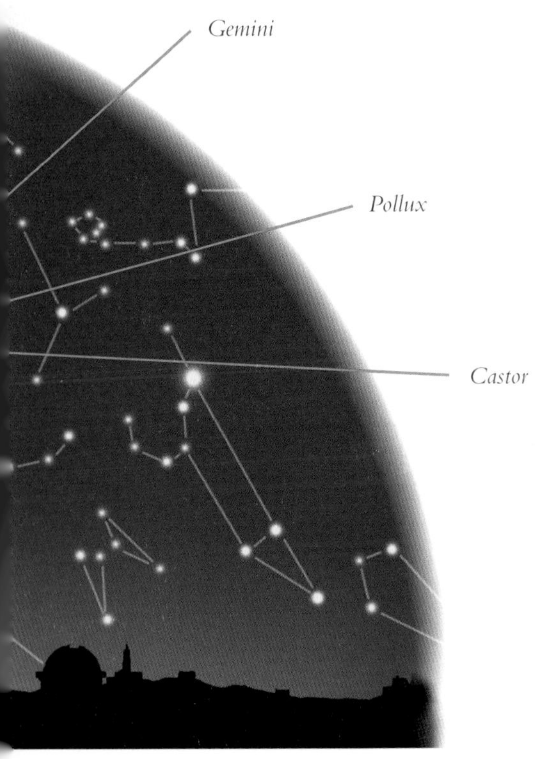

JANUARY

In the center of the sky is the constellation Auriga (the Charioteer), which is dominated by the bright yellow star Capella. This is the same type of star as our Sun but it is larger and actually comprises two stars.

To the right of Auriga is the constellation Gemini (the Twins), with its two main stars Castor and Pollux. To a visitor to the southern hemisphere, the sight of Orion (the Hunter) seemingly upside down in the night sky is quite startling!

SUMMER: LOOKING SOUTH

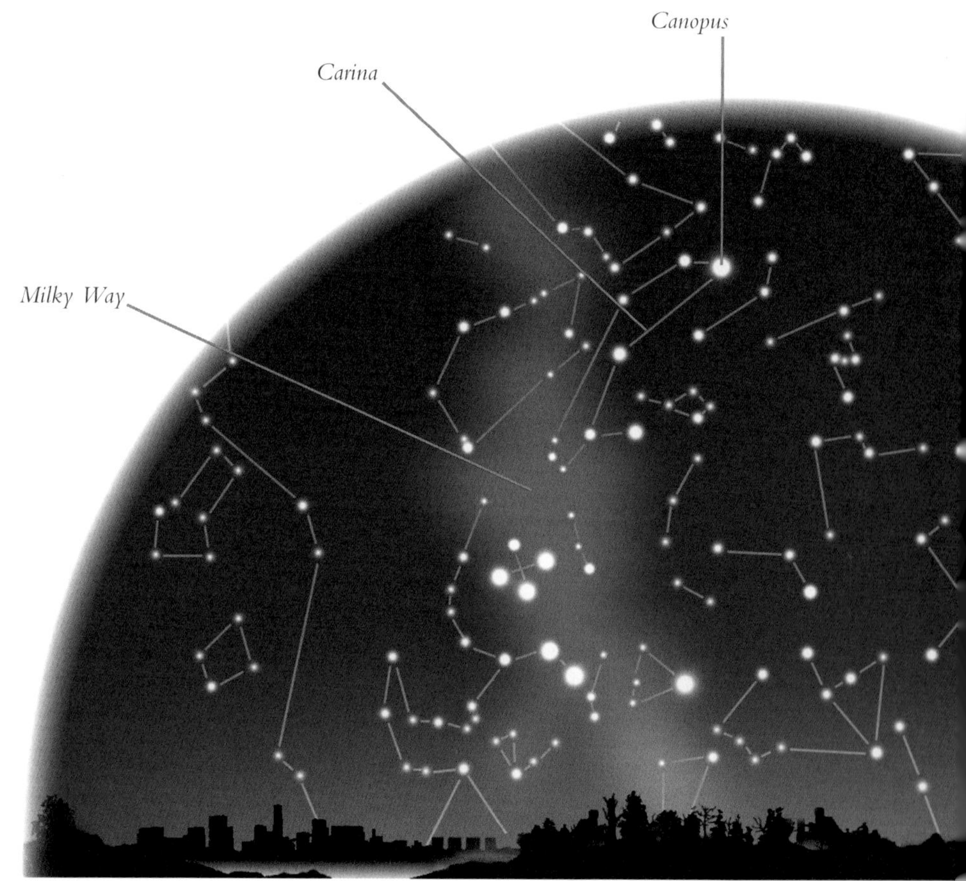

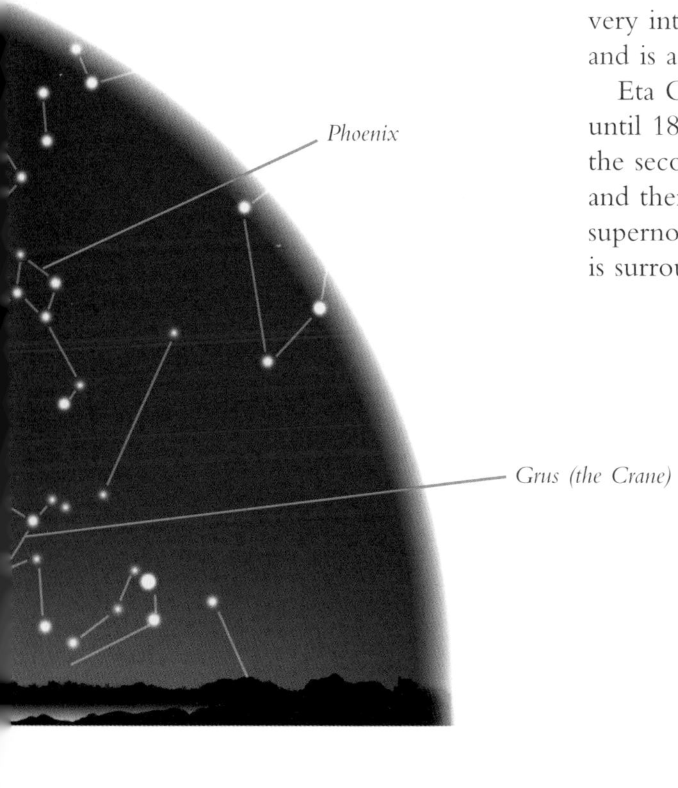

Phoenix

Grus (the Crane)

JANUARY

The constellation of Carina (the Ship's Keel) is dominated by the second brightest star in the heavens, Canopus. However, it is the rather insignificant little star just below the level of naked-eye observation that is very interesting. It is called Eta Carina and is a variable star.

Eta Carina used to be a bright star until 1843, when it surged to become the second brightest star in the sky and then faded. It is probably a supernova — an exploded star — and is surrounded by a huge shell of gas.

AUTUMN: LOOKING NORTH

APRIL

The "bent handle" of Ursa Major (the Great Bear) follows upward in an imaginary line to the stars Arcturus and Spica, two of the brightest stars in the sky.

The constellation Coma Berenices appears to be rather inconspicuous, apparently consisting of just two bright stars above Ursa Major's bent handle. However, it is the home of the Whirlpool Galaxy, one of a number of galaxies that can be seen with a telescope.

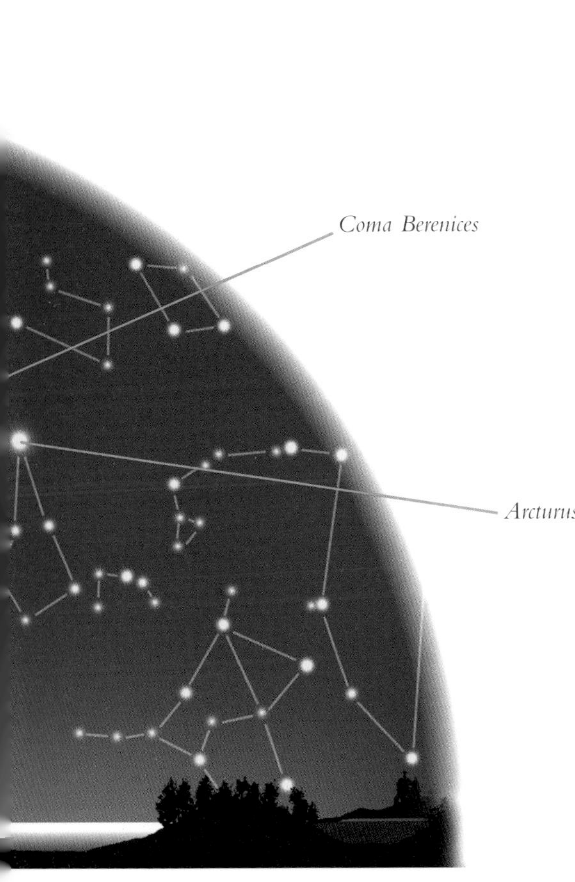

AUTUMN: LOOKING SOUTH

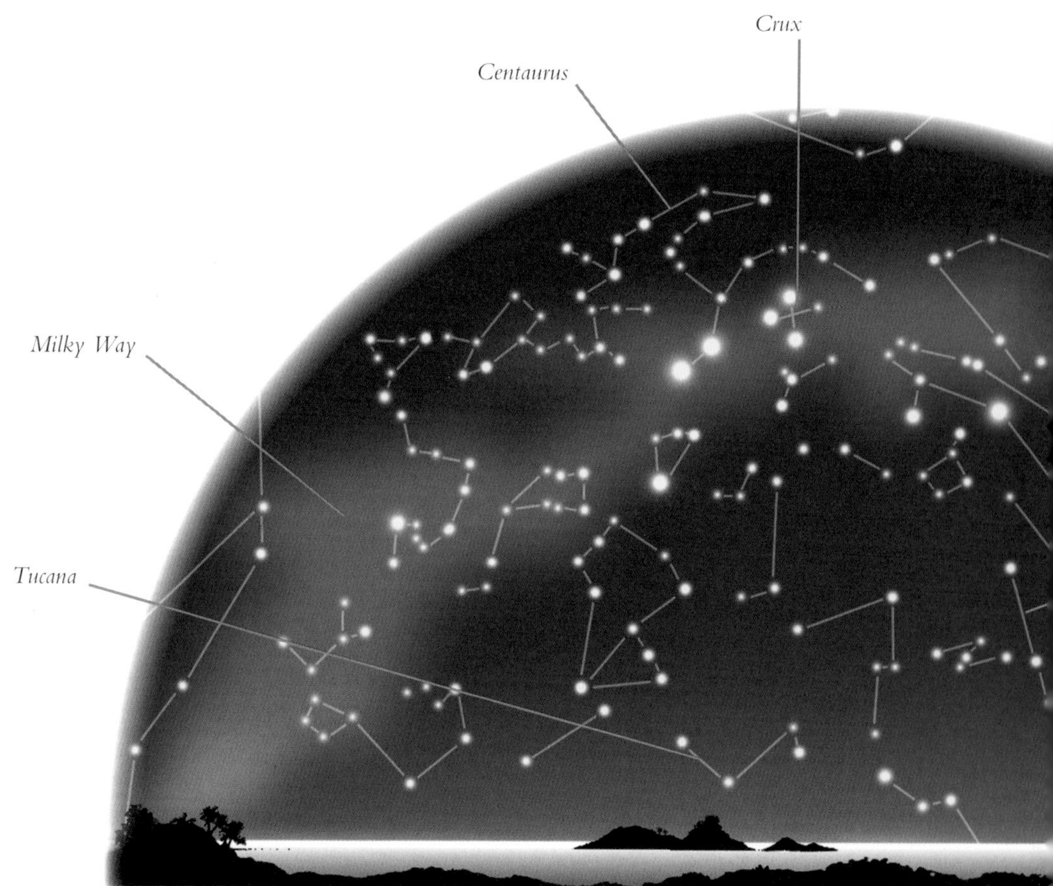

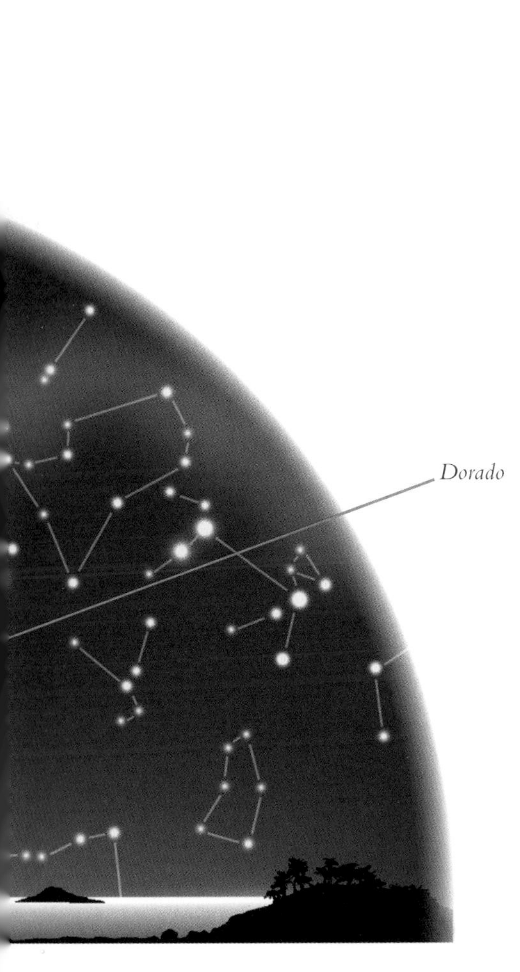

APRIL

Spring is the best time to see the constellation Crux (the Southern Cross) in its glory. Just above it, to the left, lies the constellation Centaurus (the Centaur). Its brightest star, called Alpha Centauri, is actually a system containing three stars, including a star which is invisible to the naked eye, Proxima Centauri. It lies 4.3 light-years away and is the nearest star to us other than the Sun.

The constellation Dorado (the Swordfish) contains the Large Magellanic Cloud, which is a small companion galaxy to our own Milky Way Galaxy. The Small Magellanic Cloud can be found in the constellation Tucana (the Toucan).

WINTER: LOOKING NORTH

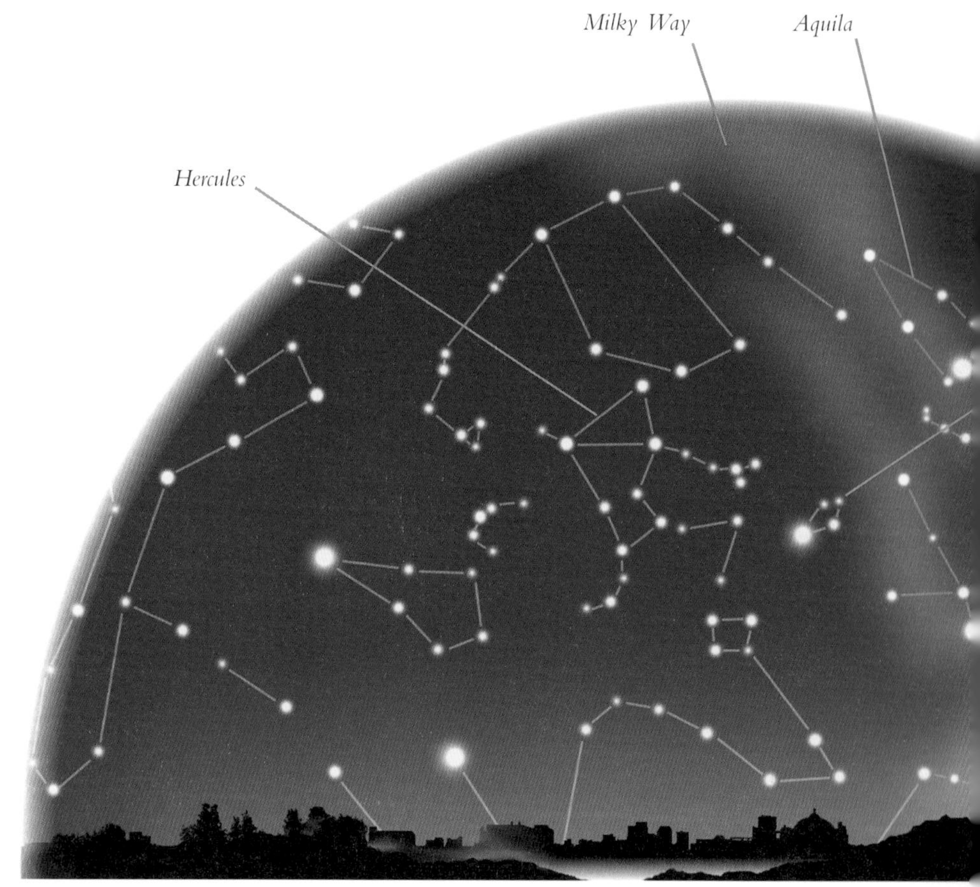

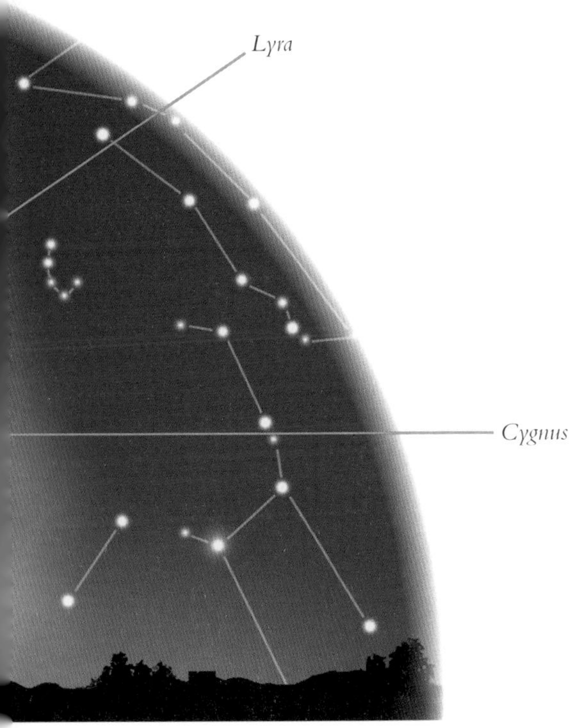

JULY

The Milky Way is a beautiful sight at this time of the year in an area dominated by three constellations: Cygnus (the Swan), Lyra (the Harp), and Aquila (the Eagle). The main stars of these constellations — Deneb, Vega, and Altair respectively — form an attractive triangle.

To the left of the triangle of stars is the constellation Hercules, which is dominated by what resembles an irregular "box." It is meant to represent the body of the well-known mythological character.

WINTER: LOOKING SOUTH

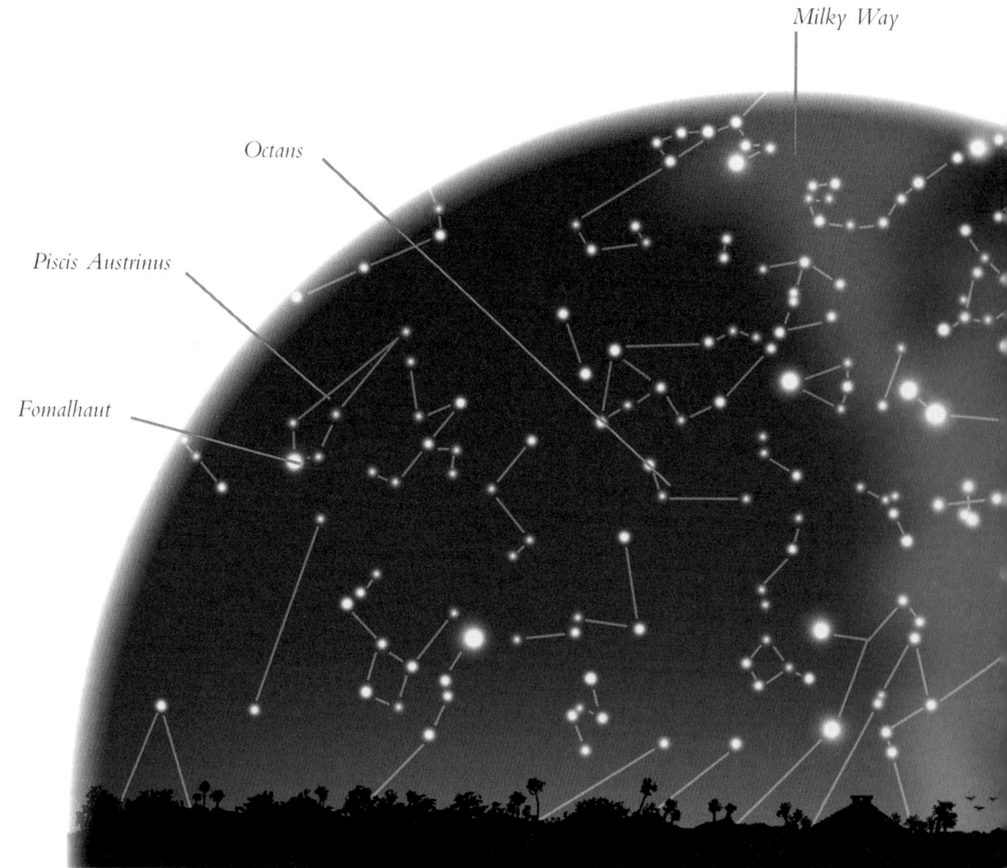

JULY

The "pole star" of the southern hemisphere can be found in the constellation Octans. It is the top right star of the constellation, which is the nearest to the south celestial pole of the sky.

The bright white star Fomalhaut, in the constellation Piscis Austrinus (Southern Fish), is 23 light-years from us but it is 11 times more luminous than our Sun.

ASTRONOMY

3000 B.C.
The Egyptians develop the science of astronomy with a basic understanding of the movements and positions of the stars.

2000 B.C.
Passages in The Bible indicate that the Earth is round and is suspended in space.

600 B.C.
About 1,400 years later, Greek philosopher Thales of Miletus is credited with the idea that the world is round.

300 B.C.
Greek astronomer Aristarchus suggests that the Earth travels around the Sun.

130 B.C.

Greek astronomer Hipparchus compiles a catalogue of 1,000 stars in the night sky.

A.D. 140
Ptolemy of Alexandria produces an encyclopedia of the knowledge of astronomy.

A.D. 1054
Chinese astronomers record the supernova that caused the Crab Nebula, which we can see today in the constellation Taurus.

A.D. 1543
Nikolai Copernicus, a Polish astronomer, says that the Earth is a planet orbiting the Sun. The age-old idea that the Earth is at the center of a Universe which revolved around it was proved to be false.

A.D. 1609
Galileo makes the first known astronomical observations using a telescope. He sees craters on the Moon, the moons of Saturn, and phases of Venus.

A.D. 1655
Huygens discovers the moon Titan orbiting Saturn and describes the true nature of the planet's ring system.

A.D. 1666
Isaac Newton establishes the basic laws of gravity. Two years later he builds the first reflecting telescope.

A.D. 1705
Edmond Halley predicts that a comet will appear in the skies in 1758. It did, and the comet was named for him.

A.D. 1781
British astronomer William Herschel discovers the first planet using a telescope. It is called Uranus.

A.D. 1801
Giuseppe Piazzi, an Italian astronomer, discovers the first asteroid, Ceres.

A.D. 1846
Basing his work on predictions from other astronomers, Johann Galle finds the planet Neptune.

A.D. 1925
Edwin Hubble uses a telescope to show that the Universe is apparently expanding, leading to the 'Big Bang' theory of how the Universe was created.

A.D. 1930
Pluto is discovered by Clyde Tombaugh.

A.D. 1931
Radio waves are detected coming from outer space. Six years later, the first radio telescope is built.

A.D. 1955
The UK's world-famous Jodrell Bank radio telescope is built.

A.D. 1990
The Hubble Space Telescope is launched into orbit. Its spectacular images revolutionize astronomy, enabling us to see 14 billion light-years into the Universe.

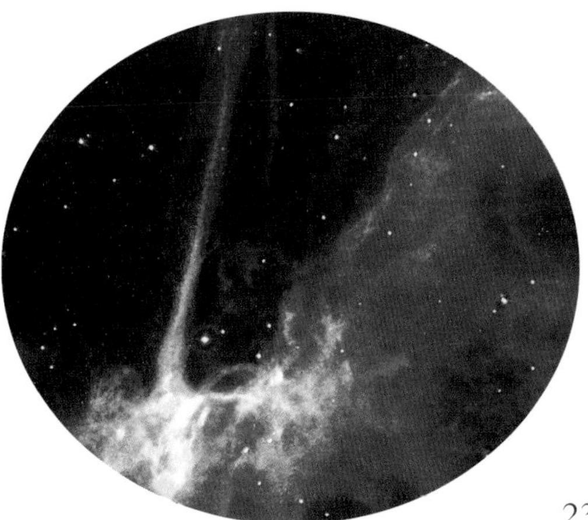

Space Exploration

1959
Luna 2 becomes the first object to contact another body in space — the Moon.

1959
Luna 3 photographs most of the far side of the Moon.

1962
Mariner 2 makes the first exploration of another planet, during a fly-by of Venus.

1964
Ranger 7 takes the first close-up images of the Moon.

1965
Mariner 4 takes the first images of the planet Mars.

1966
Luna 9 makes a soft landing on the Moon and returns the first images of its surface.

1966
Luna 10 becomes the first craft to enter orbit around the Moon.

1968
Apollo 8 carries three men around the Moon ten times.

1969.
Apollo 11 lands two men on the Moon. They complete the first Moon walk and bring back to Earth the first samples of Moon material.

1970
Venera 7 makes the first landing on Venus.

1970
The unmanned *Luna 16* returns samples of the Moon to Earth.

1971
Mariner 9 becomes the first craft to enter orbit around Mars.

1973
Pioneer 10 flies past the planet Jupiter, taking the first closeup images of the planet and its moons.

1974
Pioneer 11 flies past the planet Saturn, sending back closeup images.

1974
Mariner 10 makes the first, and so far only, explorations of Mercury during a series of flybys.

1975
The first pictures from the surface of Venus are returned by *Venera 9* and *10.*

1976
Viking 1 and *2 make* soft landings on the surface of Mars and sample its soil.

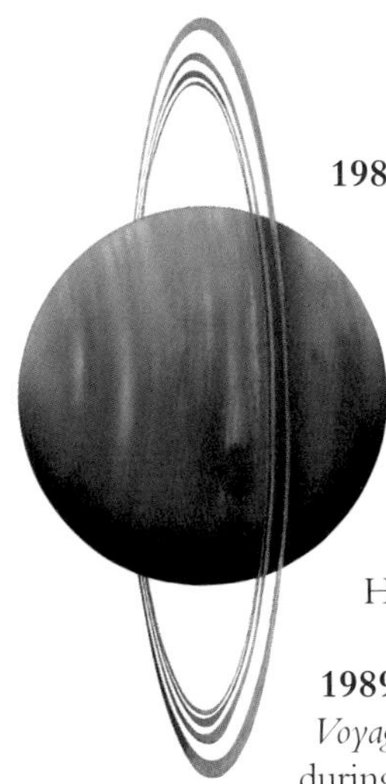

1986
Voyager 2 explores Uranus during a flyby.

1986
Giotto is one of an armada of spacecraft to explore Halley's Comet.

1989
Voyager 2 explores Neptune during a flyby.

1991
Galileo takes the first closeup look at an asteroid, Gaspra.

1995
Galileo enters orbit around Jupiter and a small probe explores its stormy atmosphere.

1997
The Mars *Pathfinder* spacecraft deploys its rover vehicle *Sojourner* after a successful soft landing on the planet's surface.

Manned Flights

1961
First manned space flights by Soviet cosmonaut Yuri Gagarin (orbital) and American astronaut Alan Shepard (sub-orbital).

1963
First woman in space is Valentina Tereshkova of the Soviet Union.

1964
Soviet Union launches three men in *Voskhod 1*.

1965
Alexei Leonov makes first walk in space from *Voskhod 2*.

1965
Gemini 6 makes first space rendezvous with *Gemini 7* which is already in space.

1966
Gemini 8 makes first space docking with *Agena* target rocket.

1967
Three *Apollo* astronauts are killed in a spacecraft fire on the launch pad.

1967
Vladimir Komarov is killed in crash landing at the end of the *Soyuz 1* mission.

1968
Apollo 8 orbits the Moon with astronauts Frank Borman, James Lovell, and Bill Anders on board.

1969
Apollo 11 makes first Moon landing, and first Moon walk is made by Neil Armstrong and Buzz Aldrin.

1970
Apollo 13 crew survives aborted Moon flight emergency.

1971
Three *Soyuz 11* cosmonauts die at the end of a record-breaking 23-day space flight. They had boarded the first space station, *Salyut 1*.

1971
First manned lunar roving vehicle is driven on the Moon by *Apollo 15* astronauts.

1972
Final Moon landing by *Apollo 17*.

1973
America launches *Skylab* space station which is later manned by three teams of three astronauts.

1974
The final *Skylab* team record an 84-day stay in space.

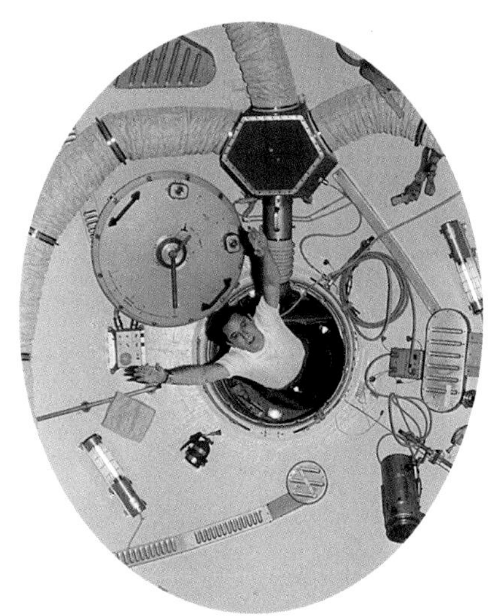

1975
American *Apollo 18* docks with Soviet *Soyuz 19* for first joint space flight.

1977
Salyut 6 cosmonauts stay 96 days in space.

1978
A Czechoslovakian cosmonaut flying a Soviet mission becomes the first non-American, non-Soviet person in space.

1980
Space endurance record increased to 185 days by *Salyut 6*'s Leonid Popov and Valeri Ryumin.

1981
NASA launches the first space shuttle, *Columbia*, with John Young and Robert Crippen aboard.

1982
Salyut 7 cosmonauts Anatoli Berezovoi and Valentin Lebedev spend over 200 days in space.

1982
Space Shuttle STS 5 mission carries four astronauts.

1983
Space Shuttle carries crews numbering five and six people for the first time.

1984
First independent space walk by Bruce McCandless operating an MMU.

1984
Russian Svetlana Savitskaya becomes the first female space-walker.

1984
Space Shuttle takes off on the 100th manned space flight.

1984
Space Shuttle carries record crew of seven.

1985
Space Shuttle carries record crew of eight.

1986
Space Shuttle *Challenger* explodes during launch, killing seven astronauts.

1986
Soviet Union launches first element of the *Mir* space station.

1987
Cosmonaut Yuri Romanenko flies a 326-day mission.

1988
Vladimir Titov and Musa Manarov complete the first space flight lasting a year.

1994
Russian Sergei Krikalev is launched aboard a US craft — the Space Shuttle.

1995
American Norman Thagard is launched by Russian rocket to board *Mir* space station.

1995
First of series of joint Space Shuttle-*Mir* missions is launched.

1998
First elements of International Space Station are joined together in space.

SATELLITE FIRSTS

1957
First artificial Earth satellite, *Sputnik 1*, from the Soviet Union.

1958
First American satellite, first science satellite, *Explorer 1*.

1958
First experimental communications satellite, *Score 1*.

1959
First military spy satellite, *Discoverer 2*.

1960
First weather satellite, *TIROS 1*.

1960
First navigation satellite, *Transit 1B*.

1960
First recovery of satellite from orbit, *Discoverer 13*.

1960
First recovery of living creatures (two dogs) from orbit, on board *Sputnik 5*.

1962
First commercial communications satellite, *Telstar 1*.

1963
First geostationary orbiting communications satellite, *Syncom 2*.

1966
First dedicated astronomical satellite, *OAO 1*.

1966
First French satellite launch, *A1*.

1967
First automatic unmanned docking in orbit, *Cosmos 186* and *188*.

1970
First Japanese satellite launch, *Ohsumi*.

1970
First Chinese satellite launch, *Tungfanghung*.

1971
First British satellite launch, *Prospero*.

1972
First Earth resources remote-sensing satellite, *Landsat 1*.

1976
First maritime mobile communications satellite, *Marisat 1*.

1980
First Indian satellite launch, *Rohini*.

1981
First European operational satellite launch by Ariane rocket, *Meteosat 2*.

1984
First satellite to be captured, repaired and redeployed, *SMM 1*.

1984
First fully commercial satellite launch, by Ariane, *Spacenet 1*.

1985
First satellites captured and returned to Earth, *Palapa* and *Westar*.

1986
First privately operated commercial remote-sensing spacecraft, *Spot 1*.

1988
First Israeli satellite launch, *Ofeq 1*.

1988
First launch and landing of Soviet Union's unmanned Space Shuttle, *Buran*.

1988
First privately operated commercial TV satellite, *Astra 1A*.

1990
First optical telescope in orbit, Hubble Space Telescope.

GLOSSARY

asteroid
A large rocky object orbiting the Sun. Often referred to as a "minor planet."

black hole
A region in space apparently formed when a dying star collapses or implodes, creating a great force of gravity that sucks material into itself. Even light cannot escape a black hole. The material disappears into an unknown abyss that may be another dimension in the Universe.

booster
A term often used for the rocket that "boosts" a payload into space.

comet
A piece of material probably left over when the Solar System was formed. It comprises rock, ice and dust and is often described as a "dirty snowball". When a comet approaches the Sun, it heats up and sheds material in the form of a tail that is lit up by the Sun.

constellation
A group of stars that make up an imaginary 'picture' in space that look like figures and objects, such as Orion the Hunter or Taurus the Bull.

docking
The joining together of two objects in space.

EVA
The abbreviation for "extravehicular activity," a term used to describe space walking outside a spacecraft.

galaxy
A huge group of millions of stars in space. Normal galaxies come in three different shapes: spiral, elliptical and irregular.

geostationary orbit
A circular orbit around the Earth's Equator at a height at which the speed of a satellite is exactly the same as the speed of the Earth's rotation. As a result, the satellite appears to be stationary in the sky.

magnitude
The apparent brightness of a star to the naked eye.

meteoroid
A relatively small rocky object that was apparently left over when the Solar System was formed. Thousands of these objects orbit the Sun and many enter the Earth's atmosphere, burning up as meteors, or "shooting stars."

Milky Way
The "cloud band" of stars that we can see in the middle of the night sky. The Milky Way is toward the center of our galaxy, which we often refer to as the Milky Way galaxy.

moon
A smaller natural object orbiting a planet, such as the Earth's Moon and the Martian moon Phobos.

nebula
A gaseous and dusty region of space in which stars are being formed. A nebula may be left over when a star dies, in which case it is called a planetary nebula.

neutron star
A tiny star that is left over when a star runs out of fuel and dies.

planet
A large body that orbits the Sun or another star.

pulsar
A rotating neutron star that sends out regular pulses of radio signals.

quasar
An unexplained object that is among the most powerful and energetic bodies in the whole Universe. A pulsar emits the same energy as 100 galaxies but it is much smaller than a single galaxy.

reentry
The moment when a satellite in orbit in space is captured by the Earth's gravity and begins to be pulled back into the Earth's atmosphere.

rendezvous
When one spacecraft approaches another in orbit, after making a series of carefully controlled engine firings to match the orbit of the other craft.

satellite
An object that orbits another object in space, such as a satellite above the Earth. The Earth and the other planets are satellites of the Sun.

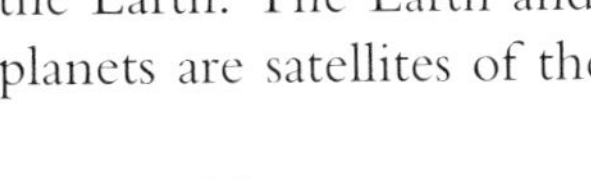

solar orbit
An orbit around the Sun.

solar system
A family of planets and other objects that orbit a star, such as the Sun.

INDEX

*Numbers in **bold** refer to main entries*

ACKNOWLEDGMENTS

The publishers wish to thank the following artists who have contributed to this book.

Julian Baker, Kuo Kang Chen, Rob Jakeway, Darrell Warner (Beehive Illustrations) Guy Smith, Janos Marffy, Peter Sarson

Photographs supplied by Genesis Photo Library and Miles Kelly Archive